INTERNATIONAL UNION OF PURE AND APPLIED CHEMISTRY

Physical and Biophysical Chemistry Division

Quantities, Units and Symbols in Physical Chemistry

Third Edition

Prepared for publication by

E. Richard Cohen	Tomislav Cvitaš	Jeremy G. Frey
Bertil Holmström	Kozo Kuchitsu	Roberto Marquardt
Ian Mills	Franco Pavese	Martin Quack
Jürgen Stohner	Herbert L. Strauss	Michio Takami

Anders J Thor

The first and second editions were prepared for publication by

Ian Mills   Tomislav Cvitaš
Klaus Homann   Nikola Kallay   Kozo Kuchitsu

IUPAC 2007

RSCPublishing

Professor E. Richard Cohen
17735, Corinthian Drive
Encino, CA 91316-3704
USA
email: ercohen@aol.com

Professor Jeremy G. Frey
University of Southampton
Department of Chemistry
Southampton, SO 17 1BJ
United Kingdom
email: j.g.frey@soton.ac.uk

Professor Kozo Kuchitsu
Tokyo University of Agriculture and Technology
Graduate School of BASE
Naka-cho, Koganei
Tokyo 184-8588
Japan
email: kuchitsu@cc.tuat.ac.jp

Professor Ian Mills
University of Reading
Department of Chemistry
Reading, RG6 6AD
United Kingdom
email: i.m.mills@rdg.ac.uk

Professor Martin Quack
ETH Zürich
Physical Chemistry
CH-8093 Zürich
Switzerland
email: Martin@Quack.ch or
quack@ir.phys.chem.ethz.ch

Professor Herbert L. Strauss
University of California
Berkeley, CA 94720-1460
USA
email: hls@berkeley.edu

Dr. Anders J Thor
Secretariat of ISO/TC 12
SIS Swedish Standards Institute
Sankt Paulsgatan 6
SE-11880 Stockholm
Sweden
email: anders.j.thor@sis.se

Professor Tom Cvitaš
University of Zagreb
Department of Chemistry
Horvatovac 102a
HR-10000 Zagreb
Croatia
email: cvitas@chem.pmf.hr

Professor Bertil Holmström
Ulveliden 15
SE-41674 Göteborg
Sweden
email: bertilh@brikks.com

Professor Roberto Marquardt
Laboratoire de Chimie Quantique
Institut de Chimie
Université Louis Pasteur
4, Rue Blaise Pascal
F-67000 Strasbourg
France
email: roberto.marquardt@chimie.u-strasbg.fr

Professor Franco Pavese
Instituto Nazionale di Ricerca Metrologica (INRIM)
strada delle Cacce 73-91
I-10135 Torino
Italia
email: F.Pavese@imgc.cnr.it

Professor Jürgen Stohner
ZHAW Zürich University of Applied Sciences
ICBC Institute of Chemistry & Biological Chemistry
Campus Reidbach T, Einsiedlerstr. 31
CH-8820 Wädenswil
Switzerland
email: sthj@zhaw.ch or just@ir.phys.chem.ethz.ch

Professor Michio Takami
3-10-8 Atago
Niiza, Saitama 352-0021
Japan
email: takamimy@d6.dion.ne.jp

ISBN: 978-0-85404-433-7

A catalogue record for this book is available from the British Library

© International Union of Pure and Applied Chemistry 2007

Reprinted 2008, 2011

Published by The Royal Society of Chemistry,
Thomas Graham House, Science Park, Milton Road,
Cambridge CB4 0WF, UK

Registered Charity Number 207890

For further information see our web site at www.rsc.org

Printed and bound in Great Britain by CPI Antony Rowe, Chippenham and Eastbourne

CONTENTS

The purpose of this manual is to improve the exchange of scientific information among the readers in different disciplines and across different nations. As the volume of scientific literature expands, each discipline has a tendency to retreat into its own jargon. This book attempts to provide a readable compilation of widely used terms and symbols from many sources together with brief understandable definitions. This Third Edition reflects the experience of the contributors with the previous editions and we are grateful for the many thoughtful comments we have received. Most of the material in this book is "standard", but a few definitions and symbols are not universally accepted. In such cases, we have attempted to list acceptable alternatives. The references list the reports from IUPAC and other sources in which some of these notational problems are discussed further. IUPAC is the acronym for International Union of Pure and Applied Chemistry.

A spectacular example of the consequences of confusion of units is provided by the loss of the United States NASA satellite, the "Mars Climate Orbiter" (MCO). The Mishap Investigation Board (Phase I Report, November 10, 1999)[1] found that the root cause for the loss of the MCO was "the failure to use metric units in the coding of the ground (based) software file". The impulse was reported in Imperial units of pounds (force)-seconds (lbf-s) rather than in the metric units of Newton (force)-seconds (N-s). This caused an error of a factor of 4.45 and threw the satellite off course.[2] We urge the users of this book always to define explicitly the terms, the units, and the symbols that they use.

This edition has been compiled in machine-readable form by Martin Quack and Jürgen Stohner. The entire text of the manual will be available on the Internet some time after the publication of the book and will be accessible via the IUPAC web site, `http://www.iupac.org`. Suggestions and comments are welcome and may be addressed in care of the

> IUPAC Secretariat
> PO Box 13757
> Research Triangle Park, NC 27709-3757, USA
> email: secretariat@iupac.org

Corrections to the manual will be listed periodically.

The book has been systematically brought up to date and new sections have been added. As in previous editions, the first chapter describes the use of quantity calculus for handling physical quantities and the general rules for the symbolism of quantities and units and includes an expanded description on the use of roman and italic fonts in scientific printing. The second chapter lists the symbols for quantities in a wide range of topics used in physical chemistry. New parts of this chapter include a section on surface structure. The third chapter describes the use of the International System of units (SI) and of a few other systems such as atomic units. Chapter 4 outlines mathematical symbols and their use in print. Chapter 5 presents the 2006 revision of the fundamental physical constants, and Chapter 6 the properties of elementary particles, elements and nuclides. Conversion of units follows in Chapter 7, together with the equations of electricity and magnetism in their various forms. Chapter 8 is entirely new and outlines the treatment of uncertainty in physical measurements. Chapter 9 lists abbreviations and acronyms. Chapter 10 provides the references, and Chapter 11, the Greek alphabet. In Chapters 12 and 13, we end with indexes. The IUPAC Periodic Table of the Elements is shown on the inside back cover and the preceding pages list frequently used conversion factors for pressure and energy units.

---

[1] The MCO report can be found at ftp://ftp.hq.nasa.gov/pub/pao/reports/1999/MCO_report.pdf.
[2] Impulse (change of momentum) means here the time-integral of the force.

Many people have contributed to this volume. The people most directly responsible are acknowledged in the Historical Introduction. Many of the members of IUPAC I.1 have continued to make active contributions long after their terms on the Commission expired. We also wish to acknowledge the members of the other Commissions of the Physical Chemistry Division: Chemical Kinetics, Colloid and Surface Chemistry, Electrochemistry, Spectroscopy, and Thermodynamics, who have each contributed to the sections of the book that concern their various interests.

We also thank all those who have contributed whom we have inadvertently missed out of these lists.

Commission on Physicochemical Symbols,   Jeremy G. Frey and Herbert L. Strauss
Terminology and Units

The *Manual of Symbols and Terminology for Physicochemical Quantities and Units* [1.a], to which this is a direct successor, was first prepared for publication on behalf of the Physical Chemistry Division of IUPAC by M. L. McGlashan in 1969, when he was chairman of the Commission on Physicochemical Symbols, Terminology and Units (I.1). He made a substantial contribution towards the objective which he described in the preface to that first edition as being "to secure clarity and precision, and wider agreement in the use of symbols, by chemists in different countries, among physicists, chemists and engineers, and by editors of scientific journals". The second edition of that manual prepared for publication by M. A. Paul in 1973 [1.b], and the third edition prepared by D. H. Whiffen in 1976 [1.c], were revisions to take account of various developments in the Système International d'Unités (International System of Units, international abbreviation SI), and other developments in terminology.

The first edition of *Quantities, Units and Symbols in Physical Chemistry* published in 1988 [2.a] was a substantially revised and extended version of the earlier manuals. The decision to embark on this project originally proposed by N. Kallay was taken at the IUPAC General Assembly at Leuven in 1981, when D. R. Lide was chairman of the Commission. The working party was established at the 1983 meeting in Lyngby, when K. Kuchitsu was chairman, and the project has received strong support throughout from all present and past members of the Commission I.1 and other Physical Chemistry Commissions, particularly D. R. Lide, D. H. Whiffen, and N. Sheppard.

The extensions included some of the material previously published in appendices [1.d−1.k]; all the newer resolutions and recommendations on units by the Conférence Générale des Poids et Mesures (CGPM); and the recommendations of the International Union of Pure and Applied Physics (IUPAP) of 1978 and of Technical Committee 12 of the International Organization for Standardization, *Quantities, units, symbols, conversion factors* (ISO/TC 12). The tables of physical quantities (Chapter 2) were extended to include defining equations and SI units for each quantity. The style was also slightly changed from being a book of rules towards a manual of advice and assistance for the day-to-day use of practicing scientists. Examples of this are the inclusion of extensive notes and explanatory text inserts in Chapter 2, the introduction to quantity calculus, and the tables of conversion factors between SI and non-SI units and equations in Chapter 7.

The second edition (1993) was a revised and extended version of the previous edition. The revisions were based on the recent resolutions of the CGPM [3]; the new recommendations by IUPAP [4]; the new international standards ISO 31 [5,6]; some recommendations published by other IUPAC commissions; and numerous comments we have received from chemists throughout the world. The revisions in the second edition were mainly carried out by Ian Mills and Tom Cvitaš with substantial input from Robert Alberty, Kozo Kuchitsu, Martin Quack as well as from other members of the IUPAC Commission on Physicochemical Symbols, Terminology and Units.

The manual has found wide acceptance in the chemical community, and various editions have been translated into Russian [2.c], Hungarian [2.d], Japanese [2.e], German [2.f], Romanian [2.g], Spanish [2.h], and Catalan [2.i]. Large parts of it have been reproduced in the 71st and subsequent editions of the *Handbook of Chemistry and Physics* published by CRC Press.

The work on revisions of the second edition started immediately after its publication and between 1995 and 1997 it was discussed to change the title to "Physical-Chemical Quantities, Units and Symbols" and to apply rather complete revisions in various parts. It was emphasized that the book covers as much the field generally called "physical chemistry" as the field called "chemical physics". Indeed we consider the larger interdisciplinary field where the boundary between physics and chemistry has largely disappeared [10]. At the same time it was decided to produce the whole book as a text file in computer readable form to allow for future access directly by computer, some time after the printed version would be available. Support for this decision came from the IUPAC secretariat in the Research Triangle Park, NC (USA) (John W. Jost). The practical work on the revisions was carried out at the ETH Zürich, while the major input on this edition came from the group of editors listed now in full on the cover. It fits with the new structure of IUPAC that these are defined as project members and not only through membership in the commission. The basic structure of this edition was finally established at a working meeting of the project members in Engelberg, Switzerland in March 1999, while further revisions were discussed at the Berlin meeting (August 1999) and thereafter. In 2001 it was decided finally to use the old title. In this edition the whole text and all tables have been revised, many chapters substantially. This work was carried out mainly at ETH Zürich, where Jürgen Stohner coordinated the various contributions and corrections from the current project group members and prepared the print-ready electronic document. Larger changes compared to previous editions concern a complete and substantial update of recently available improved constants, sections on uncertainty in physical quantities, dimensionless quantities, mathematical symbols and numerous other sections.

At the end of this historical survey we might refer also to what might be called the tradition of this manual. It is not the aim to present a list of recommendations in form of commandments. Rather we have always followed the principle that this manual should help the user in what may be called "good practice of scientific language". If there are several well established uses or conventions, these have been mentioned, giving preference to one, when this is useful, but making allowance for variety, if such variety is not harmful to clarity. In a few cases possible improvements of conventions or language are mentioned with appropriate reference, even if uncommon, but without specific recommendation. In those cases where certain common uses are deprecated, there are very strong reasons for this and the reader should follow the corresponding advice.

Zürich, 2007   Martin Quack

The membership of the Commission during the period 1963 to 2006, during which the successive editions of this manual were prepared, was as follows:

Titular members

Chairman: 1963–1967 G. Waddington (USA); 1967–1971 M.L. McGlashan (UK); 1971–1973 M.A. Paul (USA); 1973–1977 D.H. Whiffen (UK); 1977–1981 D.R. Lide Jr (USA); 1981–1985 K. Kuchitsu (Japan); 1985–1989 I.M. Mills (UK); 1989–1993 T. Cvitaš (Croatia); 1993–1999 H.L. Strauss (USA); 2000–2007 J.G. Frey (UK).

Secretary: 1963–1967 H. Brusset (France); 1967–1971 M.A. Paul (USA); 1971–1975 M. Fayard (France); 1975–1979 K.G. Weil (Germany); 1979–1983 I. Ansara (France); 1983–1985 N. Kallay (Croatia); 1985–1987 K.H. Homann (Germany); 1987–1989 T. Cvitaš (Croatia); 1989–1991 I.M. Mills (UK); 1991–1997, 2001–2005 M. Quack (Switzerland); 1997–2001 B. Holmström (Sweden).

Other titular members

1975–1983 I. Ansara (France); 1965–1969 K.V. Astachov (Russia); 1963–1971 R.G. Bates (USA); 1963–1967 H. Brusset (France); 1985–1997 T. Cvitaš (Croatia); 1963 F. Daniels (USA); 1979–1981 D.H.W. den Boer (Netherlands); 1981–1989 E.T. Denisov (Russia); 1967–1975 M. Fayard (France); 1997–2005 J. Frey (UK); 1963–1965 J.I. Gerassimov (Russia); 1991–2001 B. Holmström (Sweden); 1979–1987 K.H. Homann (Germany); 1963–1971 W. Jaenicke (Germany); 1967–1971 F. Jellinek (Netherlands); 1977–1985 N. Kallay (Croatia); 1973–1981 V. Kellö (Czechoslovakia); 1989–1997 I.V. Khudyakov (Russia); 1985–1987 W.H. Kirchhoff (USA); 1971–1979 J. Koefoed (Denmark); 1979–1987 K. Kuchitsu (Japan); 1971–1981 D.R. Lide Jr (USA); 1997–2001, 2006–   R. Marquardt (France); 1963–1971 M.L. McGlashan (UK); 1983–1991 I.M. Mills (UK); 1963–1967 M. Milone (Italy); 1967–1973 M.A. Paul (USA); 1991–1999, 2006–   F. Pavese (Italy); 1963–1967 K.J. Pedersen (Denmark); 1967–1975 A. Perez-Masiá (Spain); 1987–1997 and 2002–2005 M. Quack (Switzerland); 1971–1979 A. Schuyff (Netherlands); 1967–1970 L.G. Sillén (Sweden); 1989–1999 and 2002–2005 H.L. Strauss (USA); 1995–2001 M. Takami (Japan); 1987–1991 M. Tasumi (Japan); 1963–1967 G. Waddington (USA); 1981–1985 D.D. Wagman (USA); 1971–1979 K.G. Weil (Germany); 1971–1977 D.H. Whiffen (UK); 1963–1967 E.H. Wiebenga (Netherlands).

Associate members

1983–1991 R.A. Alberty (USA); 1983–1987 I. Ansara (France); 1979–1991 E.R. Cohen (USA); 1979–1981 E.T. Denisov (Russia); 1987–1991 G.H. Findenegg (Germany); 1987–1991 K.H. Homann (Germany); 1971–1973 W. Jaenicke (Germany); 1985–1989 N. Kallay (Croatia); 1987–1989 and 1998–1999 I.V. Khudyakov (Russia); 1979–1980 J. Koefoed (Denmark); 1987–1991 K. Kuchitsu (Japan); 1981–1983 D.R. Lide Jr (USA); 1971–1979 M.L. McGlashan (UK); 1991–1993 I.M. Mills (UK); 1973–1981 M.A. Paul (USA); 1999–2005 F. Pavese (Italy); 1975–1983 A. Perez-Masiá (Spain); 1997–1999 M. Quack (Switzerland); 1979–1987 A. Schuyff (Netherlands); 1963–1971 S. Seki (Japan); 2000–2001 H.L. Strauss (USA); 1991–1995 M. Tasumi (Japan); 1969–1977 J. Terrien (France); 1994–2001 A J Thor (Sweden); 1975–1979 L. Villena (Spain); 1967–1969 G. Waddington (USA); 1979–1983 K.G. Weil (Germany); 1977–1985 D.H. Whiffen (UK).

National representatives

Numerous national representatives have served on the commission over many years. We do not provide this long list here.

1   PHYSICAL QUANTITIES AND UNITS

1.1 PHYSICAL QUANTITIES AND QUANTITY CALCULUS

The value of a *physical quantity* Q can be expressed as the product of a *numerical value* $\{Q\}$ and a *unit* $[Q]$

$$Q = \{Q\}\,[Q] \tag{1}$$

Neither the name of the physical quantity, nor the symbol used to denote it, implies a particular choice of unit (see footnote [1], p. 4).

Physical quantities, numerical values, and units may all be manipulated by the ordinary rules of algebra. Thus we may write, for example, for the wavelength λ of one of the yellow sodium lines

$$\lambda = 5.896 \times 10^{-7}\ \mathrm{m} = 589.6\ \mathrm{nm} \tag{2}$$

where m is the symbol for the unit of length called the metre (or meter, see Sections 3.2 and 3.3, p. 86 and 87), nm is the symbol for the nanometre, and the units metre and nanometre are related by

$$1\ \mathrm{nm} = 10^{-9}\ \mathrm{m} \qquad \text{or} \qquad \mathrm{nm} = 10^{-9}\ \mathrm{m} \tag{3}$$

The equivalence of the two expressions for λ in Equation (2) follows at once when we treat the units by the rules of algebra and recognize the identity of 1 nm and 10^{-9} m in Equation (3). The wavelength may equally well be expressed in the form

$$\lambda/\mathrm{m} = 5.896 \times 10^{-7} \tag{4}$$

or

$$\lambda/\mathrm{nm} = 589.6 \tag{5}$$

It can be useful to work with variables that are defined by dividing the quantity by a particular unit. For instance, in tabulating the numerical values of physical quantities or labeling the axes of graphs, it is particularly convenient to use the quotient of a physical quantity and a unit in such a form that the values to be tabulated are numerical values, as in Equations (4) and (5).

Example

$$
\begin{aligned}
\ln(p/\mathrm{MPa}) &= a + b/T = \\
&= a + b'(10^3\ \mathrm{K}/T) \tag{6}
\end{aligned}
$$

T/K	$10^3\ \mathrm{K}/T$	p/MPa	$\ln(p/\mathrm{MPa})$
216.55	4.6179	0.5180	-0.6578
273.15	3.6610	3.4853	1.2486
304.19	3.2874	7.3815	1.9990

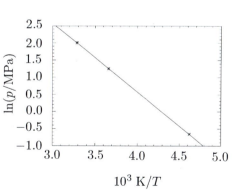

Algebraically equivalent forms may be used in place of $10^3\ \mathrm{K}/T$, such as kK/T or $10^3\ (T/\mathrm{K})^{-1}$. Equations between numerical values depend on the choice of units, whereas equations between quantities have the advantage of being independent of this choice. Therefore the use of equations between quantities should generally be preferred.

The method described here for handling physical quantities and their units is known as *quantity calculus* [11–13]. It is recommended for use throughout science and technology. The use of quantity calculus does not imply any particular choice of units; indeed one of the advantages of quantity calculus is that it makes changes between units particularly easy to follow. Further examples of the use of quantity calculus are given in Section 7.1, p. 131, which is concerned with the problems of transforming from one set of units to another.

1.2   BASE QUANTITIES AND DERIVED QUANTITIES

By convention physical quantities are organized in a dimensional system built upon seven *base quantities*, each of which is regarded as having its own dimension. These base quantities in the International System of Quantities (ISQ) on which the International System of units (SI) is based, and the principal symbols used to denote them and their dimensions are as follows:

Base quantity	Symbol for quantity	Symbol for dimension
length	l	L
mass	m	M
time	t	T
electric current	I	I
thermodynamic temperature	T	Θ
amount of substance	n	N
luminous intensity	I_v	J

All other quantities are called *derived quantities* and are regarded as having dimensions derived algebraically from the seven base quantities by multiplication and division.

Example   dimension of energy is equal to dimension of $\mathsf{M\,L^2\,T^{-2}}$
This can be written with the symbol dim for dimension (see footnote [1], below)
$\dim(E) = \dim(m \cdot l^2 \cdot t^{-2}) = \mathsf{M\,L^2\,T^{-2}}$

The quantity *amount of substance* is of special importance to chemists. Amount of substance is proportional to the number of specified elementary entities of the substance considered. The proportionality factor is the same for all substances; its reciprocal is the *Avogadro constant* (see Section 2.10, p. 47, Section 3.3, p. 88, and Chapter 5, p. 111). The SI unit of amount of substance is the mole, defined in Section 3.3, p. 88. The physical quantity "amount of substance" should no longer be called "number of moles", just as the physical quantity "mass" should not be called "number of kilograms". The name "amount of substance", sometimes also called "chemical amount", may often be usefully abbreviated to the single word "amount", particularly in such phrases as "amount concentration" (see footnote [2], below), and "amount of N_2". A possible name for international usage has been suggested: "enplethy" [10] (from Greek, similar to enthalpy and entropy).

The number and choice of base quantities is pure convention. Other quantities could be considered to be more fundamental, such as electric charge Q instead of electric current I.

$$Q = \int_{t_1}^{t_2} I \, \mathrm{d}t \tag{7}$$

However, in the ISQ, electric current is chosen as base quantity and ampere is the SI base unit. In atomic and molecular physics, the so-called *atomic units* are useful (see Section 3.9, p. 94).

---

[1] The symbol $[Q]$ was formerly used for *dimension* of Q, but this symbol is used and preferred for *unit* of Q.
[2] The Clinical Chemistry Division of IUPAC recommended that "amount-of-substance concentration" be abbreviated "substance concentration" [14].

1.3 SYMBOLS FOR PHYSICAL QUANTITIES AND UNITS [5.a]

A clear distinction should be drawn between the names and symbols for physical quantities, and the names and symbols for units. Names and symbols for many quantities are given in Chapter 2, p. 11; the symbols given there are *recommendations*. If other symbols are used they should be clearly defined. Names and symbols for units are given in Chapter 3, p. 83; the symbols for units listed there are quoted from the Bureau International des Poids et Mesures (BIPM) and are *mandatory*.

1.3.1 General rules for symbols for quantities

The symbol for a physical quantity should be a single letter (see footnote [1], below) of the Latin or Greek alphabet (see Section 1.6, p. 7). Capital or lower case letters may both be used. The letter should be printed in italic (sloping) type. When necessary the symbol may be modified by subscripts and superscripts of specified meaning. Subscripts and superscripts that are themselves symbols for physical quantities or for numbers should be printed in italic type; other subscripts and superscripts should be printed in Roman (upright) type.

Examples	C_p	for heat capacity at constant pressure
	p_i	for partial pressure of the ith substance
but	C_B	for heat capacity of substance B
	$\mu_B{}^\alpha$	for chemical potential of substance B in phase α
	E_k	for kinetic energy
	μ_r	for relative permeability
	$\Delta_r H^\ominus$	for standard reaction enthalpy
	V_m	for molar volume
	A_{10}	for decadic absorbance

The meaning of symbols for physical quantities may be further qualified by the use of one or more subscripts, or by information contained in parentheses.

Examples
$$\Delta_f S^\ominus (\mathrm{HgCl_2, cr, 25\ ^\circ C}) = -154.3\ \mathrm{J\ K^{-1}\ mol^{-1}}$$
$$\mu_i = (\partial G / \partial n_i)_{T,p,\dots,n_j,\dots;\, j \neq i} \text{ or } \mu_i = (\partial G / \partial n_i)_{T,p,n_j \neq i}$$

Vectors and matrices may be printed in bold-face italic type, e.g. $\boldsymbol{A, a}$. Tensors may be printed in bold-face italic sans serif type, e.g. S, T. Vectors may alternatively be characterized by an arrow, $\vec{A}$, $\vec{a}$ and second-rank tensors by a double arrow, $\overset{\leftrightarrow}{S}, \overset{\leftrightarrow}{T}$.

1.3.2 General rules for symbols for units

Symbols for units should be printed in Roman (upright) type. They should remain unaltered in the plural, and should not be followed by a full stop except at the end of a sentence.

Example $r = 10$ cm, not cm. or cms.

---

[1] An exception is made for certain characteristic numbers or "dimensionless quantities" used in the study of transport processes for which the internationally agreed symbols consist of two letters (see Section 2.15.1, p. 82).

Example Reynolds number, Re; another example is pH (see Sections 2.13 and 2.13.1 (viii), p. 70 and 75).

When such symbols appear as factors in a product, they should be separated from other symbols by a space, multiplication sign, or parentheses.

Symbols for units shall be printed in lower case letters, unless they are derived from a personal name when they shall begin with a capital letter. An exception is the symbol for the litre which may be either L or l, i.e. either capital or lower case (see footnote [2], below).

Examples   m (metre),   s (second),   but J (joule),   Hz (hertz)

Decimal multiples and submultiples of units may be indicated by the use of prefixes as defined in Section 3.6, p. 91.

Examples   nm (nanometre),   MHz (megahertz),   kV (kilovolt)

1.4   USE OF THE WORDS "EXTENSIVE", "INTENSIVE", "SPECIFIC", AND "MOLAR"

A quantity that is additive for independent, noninteracting subsystems is called *extensive*; examples are mass m, volume V, Gibbs energy G. A quantity that is independent of the extent of the system is called *intensive*; examples are temperature T, pressure p, chemical potential (partial molar Gibbs energy) μ.

The adjective *specific* before the name of an extensive quantity is used to mean *divided by mass*. When the symbol for the extensive quantity is a capital letter, the symbol used for the specific quantity is often the corresponding lower case letter.

Examples   volume, V, and
                     specific volume, $v = V/m = 1/\rho$ (where ρ is mass density);
                     heat capacity at constant pressure, C_p, and
                     specific heat capacity at constant pressure, $c_p = C_p/m$

ISO [5.a] and the Clinical Chemistry Division of IUPAC recommend systematic naming of physical quantities derived by division with mass, volume, area, and length by using the attributes massic or specific, volumic, areic, and lineic, respectively. In addition the Clinical Chemistry Division of IUPAC recommends the use of the attribute entitic for quantities derived by division with the number of entities [14]. Thus, for example, the specific volume could be called massic volume and the surface charge density would be areic charge.

The adjective *molar* before the name of an extensive quantity generally means *divided by amount of substance*. The subscript m on the symbol for the extensive quantity denotes the corresponding molar quantity.

Examples   volume, V molar volume, $V_{\mathrm{m}} = V/n$ (Section 2.10, p. 47)
                     enthalpy, H molar enthalpy, $H_{\mathrm{m}} = H/n$

If the name enplethy (see Section 1.2, p. 4) is accepted for "amount of substance" one can use enplethic volume instead of molar volume, for instance. The word "molar" violates the principle that the name of the quantity should not be mixed with the name of the unit (mole in this case). The use of enplethic resolves this problem. It is sometimes convenient to divide all extensive quantities by amount of substance, so that all quantities become intensive; the subscript m may then be omitted if this convention is stated and there is no risk of ambiguity. (See also the symbols recommended for partial molar quantities in Section 2.11, p. 57, and in Section 2.11.1 (iii), p. 60.)

There are a few cases where the adjective *molar* has a different meaning, namely *divided by amount-of-substance concentration*.

Examples   absorption coefficient, a
                     molar absorption coefficient, $\varepsilon = a/c$ (see Section 2.7, note 22, p. 37)
                     conductivity, κ
                     molar conductivity, $\Lambda = \kappa/c$ (see Section 2.13, p. 73)

---

[2] However, only the lower case l is used by ISO and the International Electrotechnical Commission (IEC).

1.5 PRODUCTS AND QUOTIENTS OF PHYSICAL QUANTITIES AND UNITS

Products of physical quantities may be written in any of the ways

$$a\,b \quad \text{or} \quad ab \quad \text{or} \quad a \cdot b \quad \text{or} \quad a \times b$$

and similarly quotients may be written

$$a/b \quad \text{or} \quad \frac{a}{b} \quad \text{or by writing the product of } a \text{ and } b^{-1}, \text{ e.g. } ab^{-1}$$

Examples $F = ma, \quad p = nRT/V$

Not more than one solidus (/) shall be used in the same expression unless parentheses are used to eliminate ambiguity.

Example $(a/b)/c$ or $a/(b/c)$ (in general different),   not $a/b/c$

In evaluating combinations of many factors, multiplication written without a multiplication sign takes precedence over division in the sense that a/bc is interpreted as $a/(bc)$ and not as $(a/b)c$; however, it is necessary to use parentheses to eliminate ambiguity under all circumstances, thus avoiding expressions of the kind a/bcd etc. Furthermore, $a/b + c$ is interpreted as $(a/b) + c$ and not as $a/(b + c)$. Again, the use of parentheses is recommended (required for $a/(b + c)$).

Products and quotients of units may be written in a similar way, except that the cross ($\times$) is not used as a multiplication sign between units. When a product of units is written without any multiplication sign a space shall be left between the unit symbols.

Example $1\ \mathrm{N} = 1\ \mathrm{m\ kg\ s^{-2}} = 1\ \mathrm{m \cdot kg \cdot s^{-2}} = 1\ \mathrm{m\ kg/s^2}$,   not $1\ \mathrm{mkgs^{-2}}$

1.6 THE USE OF ITALIC AND ROMAN FONTS FOR SYMBOLS IN SCIENTIFIC PUBLICATIONS

Scientific manuscripts should follow the accepted conventions concerning the use of italic and Roman fonts for symbols. An italic font is generally used for emphasis in running text, but it has a quite specific meaning when used for symbols in scientific text and equations. The following summary is intended to help in the correct use of italic fonts in preparing manuscript material.

1. The general rules concerning the use of italic (sloping) font or Roman (upright) font are presented in Section 1.3.2, p. 5 and in Section 4.1, p. 103 in relation to mathematical symbols and operators. These rules are also presented in the International Standards ISO 31 (successively being replaced by ISO/IEC 80000) [5], ISO 1000 [6], and in the SI Brochure [3].

2. The overall rule is that symbols representing physical quantities or variables are italic, but symbols representing units, mathematical constants, or labels, are roman. Sometimes there may seem to be doubt as to whether a symbol represents a quantity or has some other meaning (such as label): a good rule is that quantities, or variables, may have a range of numerical values, but labels cannot. Vectors, tensors and matrices are denoted using a bold-face (heavy) font, but they shall be italic since they are quantities.

   Examples   The Planck constant $h = 6.626\ 068\ 96(33) \times 10^{-34}$ J s.
   The electric field strength $\boldsymbol{E}$ has components $E_x, E_y,$ and E_z.
   The mass of my pen is $m = 24$ g $= 0.024$ kg.

3. The above rule applies equally to all letter symbols from both the Greek and the Latin alphabet, although some authors resist putting Greek letters into italic.

   Example   When the symbol μ is used to denote a physical quantity (such as permeability or reduced mass) it should be italic, but when it is used as a prefix in a unit such as microgram, µg, or when it is used as the symbol for the muon, µ (see paragraph 5 below), it should be roman.

4. Numbers, and labels, are roman (upright).

   Examples   The ground and first excited electronic state of the CH_2 molecule are denoted $\ldots(2a_1)^2(1b_2)^2(3a_1)^1(1b_1)^1$, $\tilde{X}\,^3B_1$, and $\ldots(2a_1)^2(1b_2)^2(3a_1)^2$, $\tilde{a}\,^1A_1$, respectively. The π-electron configuration and symmetry of the benzene molecule in its ground state are denoted: $\ldots(a_{2u})^2(e_{1g})^4$, $\tilde{X}\,^1A_{1g}$. All these symbols are labels and are roman.

5. Symbols for elements in the periodic system shall be roman. Similarly the symbols used to represent elementary particles are always roman. (See, however, paragraph 9 below for use of italic font in chemical-compound names.)

   Examples   H, He, Li, Be, B, C, N, O, F, Ne, ... for atoms; e for the electron, p for the proton, n for the neutron, µ for the muon, α for the alpha particle, etc.

6. Symbols for physical quantities are single, or exceptionally two letters of the Latin or Greek alphabet, but they are frequently supplemented with subscripts, superscripts or information in parentheses to specify further the quantity. Further symbols used in this way are either italic or roman depending on what they represent.

   Examples H denotes enthalpy, but H_m denotes molar enthalpy (m is a mnemonic label for molar, and is therefore roman). C_p and C_V denote the heat capacity at constant pressure p and volume V, respectively (note the roman m but italic p and V). The chemical potential of argon might be denoted μ_{Ar} or $\mu(Ar)$, but the chemical potential of the ith component in a mixture would be denoted μ_i, where i is italic because it is a variable subscript.

7. Symbols for mathematical operators are always roman. This applies to the symbol Δ for a difference, δ for an infinitesimal variation, d for an infinitesimal difference (in calculus), and to capital Σ and Π for summation and product signs, respectively. The symbols π (3.141 592...), e (2.718 281..., base of natural logarithms), i (square root of minus one), etc. are always roman, as are the symbols for specified functions such as log (lg for $\log_{10}$, ln for $\log_e$, or lb for $\log_2$), exp, sin, cos, tan, erf, *div*, *grad*, **rot**, etc. The particular operators **grad** and **rot** and the corresponding symbols ∇ for **grad**, $\nabla\times$ for **rot**, and $\nabla\cdot$ for *div* are printed in bold-face to indicate the vector or tensor character following [5.k]. Some of these letters, e.g. e for elementary charge, are also sometimes used to represent physical quantities; then of course they shall be italic, to distinguish them from the corresponding mathematical symbol.

   Examples $\Delta H = H(\text{final}) - H(\text{initial})$; $(\mathrm{d}p/\mathrm{d}t)$ used for the rate of change of pressure; δx used to denote an infinitesimal variation of x. But for a damped linear oscillator the amplitude F as a function of time t might be expressed by the equation $F = F_0 \exp(-\delta t)\sin(\omega t)$ where δ is the decay coefficient (SI unit: Np/s) and ω is the angular frequency (SI unit: rad/s). Note the use of roman δ for the operator in an infinitesimal variation of x, δx, but italic δ for the decay coefficient in the product δt. Note that the products δt and ωt are both dimensionless, but are described as having the unit neper (Np = 1) and radian (rad = 1), respectively.

8. The fundamental physical constants are always regarded as quantities subject to measurement (even though they are not considered to be variables) and they should accordingly always be italic. Sometimes fundamental physical constants are used as though they were units, but they are still given italic symbols. An example is the hartree, E_h (see Section 3.9.1, p. 95). However, the electronvolt, eV, the dalton, Da, or the unified atomic mass unit, u, and the astronomical unit, ua, have been recognized as units by the Comité International des Poids et Mesures (CIPM) of the BIPM and they are accordingly given Roman symbols.

Examples c_0 for the speed of light in vacuum, m_e for the electron mass, h for the Planck constant, N_A or L for the Avogadro constant, e for the elementary charge, a_0 for the Bohr radius, etc.

The electronvolt $1 \text{ eV} = e \cdot 1 \text{ V} = 1.602\ 176\ 487(40) \times 10^{-19} \text{ J}$.

9. Greek letters are used in systematic organic, inorganic, macromolecular, and biochemical nomenclature. These should be roman (upright), since they are not symbols for physical quantities. They designate the position of substitution in side chains, ligating-atom attachment and bridging mode in coordination compounds, end groups in structure-based names for macromolecules, and stereochemistry in carbohydrates and natural products. Letter symbols for elements are italic when they are locants in chemical-compound names indicating attachments to heteroatoms, e.g. O-, N-, S-, and P-. The italic symbol H denotes indicated or added hydrogen (see reference [15]).

Examples α-ethylcyclopentaneacetic acid
β-methyl-4-propylcyclohexaneethanol
tetracarbonyl($η^4$-2-methylidenepropane-1,3-diyl)chromium
α-(trichloromethyl)-ω-chloropoly(1,4-phenylenemethylene)
α-D-glucopyranose
5α-androstan-3β-ol
N-methylbenzamide
O-ethyl hexanethioate
3H-pyrrole
naphthalen-2(1H)-one

2   TABLES OF PHYSICAL QUANTITIES

The following tables contain the internationally recommended names and symbols for the physical quantities most likely to be used by chemists. Further quantities and symbols may be found in recommendations by IUPAP [4] and ISO [5].

Although authors are free to choose any symbols they wish for the quantities they discuss, provided that they define their notation and conform to the general rules indicated in Chapter 1, it is clearly an aid to scientific communication if we all generally follow a standard notation. The symbols below have been chosen to conform with current usage and to minimize conflict so far as possible. Small variations from the recommended symbols may often be desirable in particular situations, perhaps by adding or modifying subscripts or superscripts, or by the alternative use of upper or lower case. Within a limited subject area it may also be possible to simplify notation, for example by omitting qualifying subscripts or superscripts, without introducing ambiguity. The notation adopted should in any case always be defined. Major deviations from the recommended symbols should be particularly carefully defined.

The tables are arranged by subject. The five columns in each table give the name of the quantity, the recommended symbol(s), a brief definition, the symbol for the coherent SI unit (without multiple or submultiple prefixes, see Section 3.6, p. 91), and note references. When two or more symbols are recommended, commas are used to separate symbols that are equally acceptable, and symbols of second choice are put in parentheses. A semicolon is used to separate symbols of slightly different quantities. The definitions are given primarily for identification purposes and are not necessarily complete; they should be regarded as useful relations rather than formal definitions. For some of the quantities listed in this chapter, the definitions given in various IUPAC documents are collected in [16]. Useful definitions of physical quantities in physical organic chemistry can be found in [17] and those in polymer chemistry in [18]. For dimensionless quantities, a 1 is entered in the SI unit column (see Section 3.10, p. 97). Further information is added in notes, and in text inserts between the tables, as appropriate. Other symbols used are defined within the same table (not necessarily in the order of appearance) and in the notes.

2.1 SPACE AND TIME

The names and symbols recommended here are in agreement with those recommended by IUPAP [4] and ISO [5.b].

Name	Symbols	Definition	SI unit	Notes
cartesian space coordinates	$x;\ y;\ z$		m	
cylindrical coordinates	$\rho;\ \vartheta;\ z$		m, 1, m	
spherical polar coordinates	$r;\ \vartheta;\ \varphi$		m, 1, 1	
generalized coordinates	$q,\ q_i$		(varies)	
position vector	$\boldsymbol{r}$	$\boldsymbol{r} = x\boldsymbol{e}_x + y\boldsymbol{e}_y + z\boldsymbol{e}_z$	m	
length	l		m	
special symbols:				
height	h			
breadth	b			
thickness	$d,\ \delta$			
distance	d			
radius	r			
diameter	d			
path length	s			
length of arc	s			
area	$A,\ A_{\mathrm{s}},\ S$		m^2	1
volume	$V,\ (v)$		m^3	
plane angle	$\alpha,\ \beta,\ \gamma,\ \vartheta,\ \varphi,\ ...$	$\alpha = s/r$	rad, 1	2
solid angle	$\Omega,\ (\omega)$	$\Omega = A/r^2$	sr, 1	2
time, duration	t		s	
period	T	$T = t/N$	s	3
frequency	$\nu,\ f$	$\nu = 1/T$	Hz, s^{-1}	
angular frequency	ω	$\omega = 2\pi\nu$	rad s^{-1}, s^{-1}	2, 4
characteristic time interval, relaxation time, time constant	$\tau,\ T$	$\tau = \lvert \mathrm{d}t/\mathrm{d}\ln x \rvert$	s	
angular velocity	ω	$\omega = \mathrm{d}\varphi/\mathrm{d}t$	rad s^{-1}, s^{-1}	2, 5
velocity	$\boldsymbol{v},\ \boldsymbol{u},\ \boldsymbol{w},\ \boldsymbol{c},\ \dot{\boldsymbol{r}}$	$\boldsymbol{v} = \mathrm{d}\boldsymbol{r}/\mathrm{d}t$	m s^{-1}	
speed	$v,\ u,\ w,\ c$	$v = \lvert \boldsymbol{v} \rvert$	m s^{-1}	6
acceleration	$\boldsymbol{a}$	$\boldsymbol{a} = \mathrm{d}\boldsymbol{v}/\mathrm{d}t$	m s^{-2}	7

(1) An infinitesimal area may be regarded as a vector $\boldsymbol{e}_{\mathrm{n}}\mathrm{d}A$, where $\boldsymbol{e}_{\mathrm{n}}$ is the unit vector normal to the plane.

(2) The units radian (rad) and steradian (sr) for plane angle and solid angle are derived. Since they are of dimension one (i.e. dimensionless), they may be included in expressions for derived SI units if appropriate, or omitted if clarity and meaning is not lost thereby.

(3) N is the number of identical (periodic) events during the time t.

(4) The unit Hz is not to be used for angular frequency.

(5) Angular velocity can be treated as a vector, $\boldsymbol{\omega}$, perpendicular to the plane of rotation defined by $\boldsymbol{v} = \boldsymbol{\omega} \times \boldsymbol{r}$.

(6) For the speeds of light and sound the symbol c is customary.

(7) For acceleration of free fall the symbol g is used.

2.2 CLASSICAL MECHANICS

The names and symbols recommended here are in agreement with those recommended by IUPAP [4] and ISO [5.c]. Additional quantities and symbols used in acoustics can be found in [4,5.g].

Name	Symbol	Definition	SI unit	Notes
mass	m		kg	
reduced mass	μ	$\mu = m_1 m_2/(m_1 + m_2)$	kg	
density, mass density	ρ	$\rho = m/V$	kg m^{-3}	
relative density	d	$d = \rho/\rho^{\ominus}$	1	1
surface density	ρ_A, ρ_S	$\rho_A = m/A$	kg m^{-2}	
specific volume	v	$v = V/m = 1/\rho$	m^3 kg^{-1}	
momentum	$\boldsymbol{p}$	$\boldsymbol{p} = m\boldsymbol{v}$	kg m s^{-1}	
angular momentum	$\boldsymbol{L}$	$\boldsymbol{L} = \boldsymbol{r} \times \boldsymbol{p}$	J s	2
moment of inertia	I, J	$I = \sum_i m_i r_i^2$	kg m^2	3
force	$\boldsymbol{F}$	$\boldsymbol{F} = \mathrm{d}\boldsymbol{p}/\mathrm{d}t = m\boldsymbol{a}$	N	
moment of force, torque	$\boldsymbol{M}, (\boldsymbol{T})$	$\boldsymbol{M} = \boldsymbol{r} \times \boldsymbol{F}$	N m	
energy	E		J	
potential energy	E_{p}, V, Φ	$E_{\mathrm{p}} = -\int \boldsymbol{F} \cdot \mathrm{d}\boldsymbol{r}$	J	4
kinetic energy	E_{k}, T, K	$E_{\mathrm{k}} = (1/2)mv^2$	J	
work	W, A, w	$W = \int \boldsymbol{F} \cdot \mathrm{d}\boldsymbol{r}$	J	
power	P	$P = \boldsymbol{F} \cdot \boldsymbol{v} = \mathrm{d}W/\mathrm{d}t$	W	
generalized coordinate	q		(varies)	
generalized momentum	p		(varies)	
Lagrange function	L	$L(q, \dot{q}) = T(q, \dot{q}) - V(q)$	J	
Hamilton function	H	$H(q, p) = \sum_i p_i \dot{q}_i - L(q, \dot{q})$	J	
action	S	$S = \int L \, \mathrm{d}t$	J s	5
pressure	$p, (P)$	$p = F/A$	Pa, N m^{-2}	
surface tension	γ, σ	$\gamma = \mathrm{d}W/\mathrm{d}A$	N m^{-1}, J m^{-2}	
weight	$G, (W, P)$	$G = mg$	N	
gravitational constant	G	$F = Gm_1 m_2/r^2$	N m^2 kg^{-2}	

(1) Usually $\rho^{\ominus} = \rho(\mathrm{H_2O}, 4\,°\mathrm{C})$.

(2) Other symbols are customary in atomic and molecular spectroscopy (see Section 2.6, p. 25).

(3) In general $\boldsymbol{I}$ is a tensor quantity: $I_{\alpha\alpha} = \sum_i m_i \left(\beta_i^2 + \gamma_i^2\right)$, and $I_{\alpha\beta} = -\sum_i m_i \alpha_i \beta_i$ if $\alpha \neq \beta$, where α, β, γ is a permutation of x, y, z. For a continuous mass distribution the sums are replaced by integrals.

(4) Strictly speaking, only potential energy differences have physical significance, thus the integral is to be interpreted as a definite integral, for instance

$$E_{\mathrm{p}}(r_1, r_2) = -\int_{r_1}^{r_2} \boldsymbol{F} \cdot \mathrm{d}\boldsymbol{r}$$

or possibly with the upper limit infinity

$$E_{\mathrm{p}}(r) = -\int_{r}^{\infty} \boldsymbol{F} \cdot \mathrm{d}\boldsymbol{r}$$

(5) Action is the time integral over the Lagrange function L, which is equivalent to $\int p \, \mathrm{d}q - \int H \, \mathrm{d}t$ (see [19]).

14

Name	Symbol	Definition	SI unit	Notes
normal stress	σ	$\sigma = F/A$	Pa	6
shear stress	τ	$\tau = F/A$	Pa	6
linear strain, relative elongation	ε, e	$\varepsilon = \Delta l / l$	1	
modulus of elasticity, Young's modulus	E	$E = \sigma / \varepsilon$	Pa	6
shear strain	γ	$\gamma = \Delta x / d$	1	6, 7
shear modulus, Coulomb's modulus	G	$G = \tau / \gamma$	Pa	6
volume strain, bulk strain	ϑ	$\vartheta = \Delta V / V_0$	1	6
bulk modulus, compression modulus	K	$K = -V_0 \, (\mathrm{d}p/\mathrm{d}V)$	Pa	6
viscosity, dynamic viscosity	$\eta, (\mu)$	$\tau_{xz} = \eta \, (\mathrm{d}v_x/\mathrm{d}z)$	Pa s	
fluidity	φ	$\varphi = 1/\eta$	m kg^{-1} s	
kinematic viscosity	ν	$\nu = \eta / \rho$	m^2 s^{-1}	
dynamic friction factor	$\mu, (f)$	$F_{\mathrm{frict}} = \mu F_{\mathrm{norm}}$	1	
sound energy flux	P, P_{a}	$P = \mathrm{d}E/\mathrm{d}t$	W	
acoustic factors,				
reflection	ρ	$\rho = P_{\mathrm{r}}/P_0$	1	8
absorption	$\alpha_{\mathrm{a}}, (\alpha)$	$\alpha_{\mathrm{a}} = 1 - \rho$	1	9
transmission	τ	$\tau = P_{\mathrm{tr}}/P_0$	1	8
dissipation	δ	$\delta = \alpha_{\mathrm{a}} - \tau$	1	

(6) In general these can be tensor quantities.

(7) d is the distance between the layers displaced by Δx.

(8) P_0 is the incident sound energy flux, P_{r} the reflected flux and P_{tr} the transmitted flux.

(9) This definition is special to acoustics and is different from the usage in radiation, where the absorption factor corresponds to the acoustic dissipation factor.

2.3 ELECTRICITY AND MAGNETISM

The names and symbols recommended here are in agreement with those recommended by IUPAP [4] and ISO [5.e].

Name	Symbol	Definition	SI unit	Notes
electric current	I, i		A	1
electric current density	$\boldsymbol{j}, \boldsymbol{J}$	$I = \int \boldsymbol{j} \cdot \boldsymbol{e}_\text{n}\, \mathrm{d}A$	A m^{-2}	2
electric charge, quantity of electricity	Q	$Q = \int I\, \mathrm{d}t$	C	1
charge density	ρ	$\rho = Q/V$	C m^{-3}	
surface density of charge	σ	$\sigma = Q/A$	C m^{-2}	
electric potential	V, ϕ	$V = \mathrm{d}W/\mathrm{d}Q$	V, J C^{-1}	
electric potential difference, electric tension	$U, \Delta V, \Delta \phi$	$U = V_2 - V_1$	V	
electromotive force	$E,$ $(E_\text{mf}, E_\text{MK})$	$E = \oint (\boldsymbol{F}/Q) \cdot \mathrm{d}\boldsymbol{r}$	V	3
electric field strength	$\boldsymbol{E}$	$\boldsymbol{E} = \boldsymbol{F}/Q = -\boldsymbol{\nabla}V$	V m^{-1}	
electric flux	Ψ	$\Psi = \int \boldsymbol{D} \cdot \boldsymbol{e}_\text{n}\, \mathrm{d}A$	C	2
electric displacement	$\boldsymbol{D}$	$\boldsymbol{\nabla} \cdot \boldsymbol{D} = \rho$	C m^{-2}	
capacitance	C	$C = Q/U$	F, C V^{-1}	
permittivity	ε	$\boldsymbol{D} = \varepsilon \boldsymbol{E}$	F m^{-1}	4
electric constant, permittivity of vacuum	ε_0	$\varepsilon_0 = \mu_0^{-1} c_0^{-2}$	F m^{-1}	5
relative permittivity	ε_r	$\varepsilon_\text{r} = \varepsilon/\varepsilon_0$	1	6
dielectric polarization, electric polarization (electric dipole moment per volume)	$\boldsymbol{P}$	$\boldsymbol{P} = \boldsymbol{D} - \varepsilon_0 \boldsymbol{E}$	C m^{-2}	
electric susceptibility	χ_e	$\chi_\text{e} = \varepsilon_\text{r} - 1$	1	
1st hyper-susceptibility	$\chi_\text{e}^{(2)}$	$\chi_\text{e}^{(2)} = \varepsilon_0^{-1} \left(\partial^2 P / \partial E^2\right)$	C m J^{-1}, m V^{-1}	7
2nd hyper-susceptibility	$\chi_\text{e}^{(3)}$	$\chi_\text{e}^{(3)} = \varepsilon_0^{-1} \left(\partial^3 P / \partial E^3\right)$	C^2 m^2 J^{-2}, m^2 V^{-2}	7

(1) The electric current I is a base quantity in ISQ.

(2) $\boldsymbol{e}_\text{n}\mathrm{d}A$ is a vector element of area (see Section 2.1, note 1, p. 13).

(3) The name electromotive force and the symbol emf are no longer recommended, since an electric potential difference is not a force (see Section 2.13, note 14, p. 71).

(4) ε can be a second-rank tensor.

(5) c_0 is the speed of light in vacuum.

(6) This quantity was formerly called dielectric constant.

(7) The hyper-susceptibilities are the coefficients of the non-linear terms in the expansion of the magnitude P of the dielectric polarization $\boldsymbol{P}$ in powers of the electric field strength $\boldsymbol{E}$, quite related to the expansion of the dipole moment vector described in Section 2.5, note 17, p. 24. In isotropic media, the expansion of the component i of the dielectric polarization is given by

$$P_i = \varepsilon_0 [\chi_\text{e}^{(1)} E_i + (1/2) \chi_\text{e}^{(2)} E_i^2 + (1/6) \chi_\text{e}^{(3)} E_i^3 + \cdots]$$

where E_i is the i-th component of the electric field strength, and $\chi_\text{e}^{(1)}$ is the usual electric susceptibility χ_e, equal to $\varepsilon_\text{r} - 1$ in the absence of higher terms. In anisotropic media, $\chi_\text{e}^{(1)}$, $\chi_\text{e}^{(2)}$, and $\chi_\text{e}^{(3)}$ are tensors of rank 2, 3, and 4. For an isotropic medium (such as a liquid), or for a crystal with a centrosymmetric unit cell, $\chi_\text{e}^{(2)}$ is zero by symmetry. These quantities are macroscopic

Name	Symbol	Definition	SI unit	Notes
electric dipole moment	$\boldsymbol{p}, \boldsymbol{\mu}$	$\boldsymbol{p} = \sum_i Q_i \boldsymbol{r}_i$	C m	8
magnetic flux density, magnetic induction	$\boldsymbol{B}$	$\boldsymbol{F} = Q\boldsymbol{v} \times \boldsymbol{B}$	T	9
magnetic flux	Φ	$\Phi = \int \boldsymbol{B} \cdot \boldsymbol{e}_n \, \mathrm{d}A$	Wb	2
magnetic field strength, magnetizing field strength	$\boldsymbol{H}$	$\nabla \times \boldsymbol{H} = \boldsymbol{j}$	A m^{-1}	
permeability	μ	$\boldsymbol{B} = \mu \boldsymbol{H}$	N A^{-2}, H m^{-1}	10
magnetic constant, permeability of vacuum	μ_0	$\mu_0 = 4\pi \times 10^{-7}$ H m^{-1}	H m^{-1}	
relative permeability	μ_r	$\mu_\mathrm{r} = \mu/\mu_0$	1	
magnetization (magnetic dipole moment per volume)	$\boldsymbol{M}$	$\boldsymbol{M} = \boldsymbol{B}/\mu_0 - \boldsymbol{H}$	A m^{-1}	
magnetic susceptibility	$\chi, \kappa, (\chi_\mathrm{m})$	$\chi = \mu_\mathrm{r} - 1$	1	11
molar magnetic susceptibility	χ_m	$\chi_\mathrm{m} = V_\mathrm{m}\chi$	m^3 mol^{-1}	
magnetic dipole moment	$\boldsymbol{m}, \boldsymbol{\mu}$	$E = -\boldsymbol{m} \cdot \boldsymbol{B}$	A m^2, J T^{-1}	
electric resistance	R	$R = U/I$	Ω	12
conductance	G	$G = 1/R$	S	12
loss angle	δ	$\delta = \varphi_U - \varphi_I$	rad	13
reactance	X	$X = (U/I)\sin\delta$	Ω	
impedance, (complex impedance)	Z	$Z = R + \mathrm{i}X$	Ω	
admittance, (complex admittance)	Y	$Y = 1/Z$	S	
susceptance	B	$Y = G + \mathrm{i}B$	S	
resistivity	ρ	$\boldsymbol{E} = \rho\boldsymbol{j}$	Ω m	14
conductivity	κ, γ, σ	$\boldsymbol{j} = \kappa\boldsymbol{E}$	S m^{-1}	14, 15
self-inductance	L	$E = -L(\mathrm{d}I/\mathrm{d}t)$	H, V s A^{-1}	
mutual inductance	M, L_{12}	$E_1 = -L_{12}(\mathrm{d}I_2/\mathrm{d}t)$	H, V s A^{-1}	
magnetic vector potential	$\boldsymbol{A}$	$\boldsymbol{B} = \nabla \times \boldsymbol{A}$	Wb m^{-1}	
Poynting vector	$\boldsymbol{S}$	$\boldsymbol{S} = \boldsymbol{E} \times \boldsymbol{H}$	W m^{-2}	16

(7) (continued) properties and characterize a dielectric medium in the same way that the microscopic quantities polarizability (α) and hyper-polarizabilities (β, γ) characterize a molecule. For a homogeneous, saturated, isotropic dielectric medium of molar volume V_m one has $\alpha_\mathrm{m} = \varepsilon_0 \chi_\mathrm{e} V_\mathrm{m}$, where $\alpha_\mathrm{m} = N_\mathrm{A}\alpha$ is the molar polarizability (Section 2.5, note 17, p. 24 and Section 2.7.2, p. 40).

(8) When a dipole is composed of two point charges Q and $-Q$ separated by a distance r, the direction of the dipole vector is taken to be from the negative to the positive charge. The opposite convention is sometimes used, but is to be discouraged. The dipole moment of an ion depends on the choice of the origin.

(9) This quantity should not be called "magnetic field".

(10) μ is a second-rank tensor in anisotropic materials.

(11) The symbol χ_m is sometimes used for magnetic susceptibility, but it should be reserved for molar magnetic susceptibility.

(12) In a material with reactance $R = (U/I)\cos\delta$, and $G = R/(R^2 + X^2)$.

(13) φ_I and φ_U are the phases of current and potential difference.

(14) This quantity is a tensor in anisotropic materials.

(15) ISO gives only γ and σ, but not κ.

(16) This quantity is also called the Poynting-Umov vector.

2.4 QUANTUM MECHANICS AND QUANTUM CHEMISTRY

The names and symbols for quantities used in quantum mechanics and recommended here are in agreement with those recommended by IUPAP [4]. The names and symbols for quantities used mainly in the field of quantum chemistry have been chosen on the basis of the current practice in the field. Guidelines for the presentation of methodological choices in publications of computational results have been presented [20]. A list of acronyms used in theoretical chemistry has been published by IUPAC [21]; see also Chapter 9, p. 155.

Name	Symbol	Definition	SI unit	Notes
momentum operator	$\widehat{p}$	$\widehat{p} = -i\hbar\,\boldsymbol{\nabla}$	J s m^{-1}	1
kinetic energy operator	$\widehat{T}$	$\widehat{T} = -(\hbar^2/2m)\boldsymbol{\nabla}^2$	J	1
hamiltonian operator, hamiltonian	$\widehat{H}$	$\widehat{H} = \widehat{T} + \widehat{V}$	J	1
wavefunction, state function	Ψ, ψ, ϕ	$\widehat{H}\psi = E\psi$	(m$^{-3/2}$)	2, 3
hydrogen-like wavefunction	$\psi_{nlm}(r,\theta,\phi)$	$\psi_{nlm} = R_{nl}(r)Y_{lm}(\theta,\phi)$	(m$^{-3/2}$)	3
spherical harmonic function	$Y_{lm}(\theta,\phi)$	$Y_{lm} = N_{l|m|}P_l^{|m|}(\cos\theta)e^{im\phi}$	1	4
probability density	P	$P = \psi^*\psi$	(m^{-3})	3, 5
charge density of electrons	ρ	$\rho = -eP$	(C m^{-3})	3, 5, 6
probability current density, probability flux	S	$S = -(i\hbar/2m)\times$ $(\psi^*\boldsymbol{\nabla}\psi - \psi\boldsymbol{\nabla}\psi^*)$	(m^{-2}s^{-1})	3
electric current density of electrons	j	$j = -eS$	(A m^{-2})	3, 6
integration element	$d\tau$	$d\tau = dx\,dy\,dz$	(varies)	
matrix element of operator $\widehat{A}$	$A_{ij}, \langle i|A|j\rangle$	$A_{ij} = \int \psi_i^*\widehat{A}\psi_j d\tau$	(varies)	7
expectation value of operator $\widehat{A}$	$\langle A\rangle, \overline{A}$	$\langle A\rangle = \int \psi^*\widehat{A}\psi d\tau$	(varies)	7

(1) The circumflex (or "hat"), $\widehat{}$, serves to distinguish an operator from an algebraic quantity. This definition applies to a coordinate representation, where $\boldsymbol{\nabla}$ denotes the nabla operator (see Section 4.2, p. 107).

(2) Capital and lower case ψ are commonly used for the time-dependent function $\Psi(x,t)$ and the amplitude function $\psi(x)$ respectively. Thus for a stationary state $\Psi(x,t) = \psi(x)\exp(-iEt/\hbar)$.

(3) For the normalized wavefunction of a single particle in three-dimensional space the appropriate SI unit is given in parentheses. Results in quantum chemistry, however, are commonly expressed in terms of atomic units (see Section 3.9.2, p. 95 and Section 7.3 (iv), p. 145; and reference [22]). If distances, energies, angular momenta, charges and masses are all expressed as dimensionless ratios r/a_0, E/E_h, etc., then all quantities are dimensionless.

(4) $P_l^{|m|}$ denotes the associated Legendre function of degree l and order $|m|$. $N_{l|m|}$ is a normalization factor.

(5) ψ^* is the complex conjugate of ψ. For an anti-symmetrized n electron wavefunction $\Psi(r_1,\cdots,r_n)$, the total probability density of electrons is $\int_2\cdots\int_n \Psi^*\Psi\,d\tau_2\cdots d\tau_n$, where the integration extends over the coordinates of all electrons but one.

(6) $-e$ is the charge of an electron.

(7) The unit is the same as for the physical quantity A that the operator represents.

Name	Symbol	Definition	SI unit	Notes
hermitian conjugate of operator $\widehat{A}$	$\widehat{A}^\dagger$	$\left(A^\dagger\right)_{ij} = \left(A_{ji}\right)^*$	(varies)	7
commutator of $\widehat{A}$ and $\widehat{B}$	$[\widehat{A},\widehat{B}]$, $[\widehat{A},\widehat{B}]_-$	$[\widehat{A},\widehat{B}] = \widehat{A}\widehat{B} - \widehat{B}\widehat{A}$	(varies)	8
anticommutator of $\widehat{A}$ and $\widehat{B}$	$[\widehat{A},\widehat{B}]_+$	$[\widehat{A},\widehat{B}]_+ = \widehat{A}\widehat{B} + \widehat{B}\widehat{A}$	(varies)	8
angular momentum operators	see SPECTROSCOPY, Section 2.6.1, p. 30			
spin wavefunction	α; β		1	9

Hückel molecular orbital theory (HMO)

Name	Symbol	Definition	SI unit	Notes
atomic-orbital basis function	χ_r		$m^{-3/2}$	3
molecular orbital	ϕ_i	$\phi_i = \sum_r \chi_r c_{ri}$	$m^{-3/2}$	3, 10
coulomb integral	H_{rr}, α_r	$H_{rr} = \int \chi_r{}^* \widehat{H} \chi_r \mathrm{d}\tau$	J	3, 10, 11
resonance integral	H_{rs}, β_{rs}	$H_{rs} = \int \chi_r{}^* \widehat{H} \chi_s \mathrm{d}\tau$	J	3, 10, 12
energy parameter	x	$-x = (\alpha - E)/\beta$	1	13
overlap integral	S_{rs}, S	$S_{rs} = \int \chi_r{}^* \chi_s \mathrm{d}\tau$	1	10
charge order	q_r	$q_r = \sum_{i=1}^{n} b_i c_{ri}{}^2$	1	14, 15
bond order	p_{rs}	$p_{rs} = \sum_{i=1}^{n} b_i c_{ri} c_{si}$	1	15, 16

(8) The unit is the same as for the product of the physical quantities A and B.

(9) The spin wavefunctions of a single electron, α and β, are defined by the matrix elements of the z component of the spin angular momentum, $\widehat{s}_z$, by the relations $\langle \alpha | \widehat{s}_z | \alpha \rangle = +(1/2)$, $\langle \beta | \widehat{s}_z | \beta \rangle = -(1/2)$, $\langle \beta | \widehat{s}_z | \alpha \rangle = \langle \alpha | \widehat{s}_z | \beta \rangle = 0$ in units of $\hbar$. The total electron spin wavefunctions of an atom with many electrons are denoted by Greek letters α, β, γ, etc. according to the value of $\sum m_s$, starting from the greatest down to the least.

(10) $\widehat{H}$ is an effective hamiltonian for a single electron, i and j label the molecular orbitals, and r and s label the atomic orbitals. In Hückel MO theory, H_{rs} is taken to be non-zero only for bonded pairs of atoms r and s, and all S_{rs} are assumed to be zero for $r \neq s$.

(11) Note that the name "coulomb integral" has a different meaning in HMO theory (where it refers to the energy of the orbital χ_r in the field of the nuclei) from Hartree-Fock theory discussed below (where it refers to a two-electron repulsion integral).

(12) This expression describes a bonding interaction between atomic orbitals r and s. For an anti-bonding interaction, the corresponding resonance integral is given by the negative value of the resonance integral for the bonding interaction.

(13) In the simplest application of Hückel theory to the π electrons of planar conjugated hydrocarbons, α is taken to be the same for all carbon atoms, and β to be the same for all bonded pairs of carbon atoms; it is then customary to write the Hückel secular determinant in terms of the dimensionless parameter x.

(14) $-eq_r$ is the electronic charge on atom r. q_r specifies the contribution of all n π electrons to the total charge at center r, with $\sum q_r = n$.

(15) b_i gives the number of electrons which occupy a given orbital energy level ε_i; for non-degenerate orbitals, b_i can take the values 0, 1, or 2.

(16) p_{rs} is the bond order between atoms r and s.

2.4.1 Ab initio Hartree-Fock self-consistent field theory (ab initio SCF)

Results in quantum chemistry are typically expressed in atomic units (see Section 3.9.1, p. 94 and Section 7.3 (iv), p. 145). In the remaining tables of this section all lengths, energies, masses, charges and angular momenta are expressed as dimensionless ratios to the corresponding atomic units, a_0, E_h, m_e, e and $\hbar$ respectively. Thus all quantities become dimensionless, and the SI unit column is therefore omitted.

Name	Symbol	Definition	Notes
molecular orbital	$\phi_i(\mu)$		17
molecular spin orbital	$\phi_i(\mu)\,\alpha(\mu)$; $\phi_i(\mu)\,\beta(\mu)$		17
total wavefunction	Ψ	$\Psi = (n!)^{-1/2}\,\|\phi_i(\mu)\|$	17, 18
core hamiltonian of a single electron	$\widehat{H}_\mu^{\,\text{core}}$	$\widehat{H}_\mu = -(1/2)\nabla_\mu^2 - \sum_A Z_A/r_{\mu A}$	17, 19
one-electron integrals:			
expectation value of the core hamiltonian	H_{ii}	$H_{ii} = \int \phi_i^*(1)\widehat{H}_1^{\,\text{core}}\phi_i(1)\mathrm{d}\tau_1$	17, 19
two-electron repulsion integrals:			
coulomb integral	J_{ij}	$J_{ij} = \iint \phi_i^*(1)\phi_j^*(2)\frac{1}{r_{12}}\phi_i(1)\phi_j(2)\mathrm{d}\tau_1\mathrm{d}\tau_2$	17, 20
exchange integral	K_{ij}	$K_{ij} = \iint \phi_i^*(1)\phi_j^*(2)\frac{1}{r_{12}}\phi_j(1)\phi_i(2)\mathrm{d}\tau_1\mathrm{d}\tau_2$	17, 20
one-electron orbital			
energy	ε_i	$\varepsilon_i = H_{ii} + \sum_j \left(2J_{ij} - K_{ij}\right)$	17, 21
total electronic energy	E	$E = 2\sum_i H_{ii} + \sum_i \sum_j \left(2J_{ij} - K_{ij}\right)$ $= \sum_i \left(\varepsilon_i + H_{ii}\right)$	17, 21, 22
coulomb operator	$\widehat{J}_i$	$\widehat{J}_i\phi_j(2) = \left\langle \phi_i(1)\left|\frac{1}{r_{12}}\right|\phi_i(1)\right\rangle \phi_j(2)$	17
exchange operator	$\widehat{K}_i$	$\widehat{K}_i\phi_j(2) = \left\langle \phi_i(1)\left|\frac{1}{r_{12}}\right|\phi_j(1)\right\rangle \phi_i(2)$	17
Fock operator	$\widehat{F}$	$\widehat{F} = \widehat{H}^{\,\text{core}} + \sum_i \left(2\widehat{J}_i - \widehat{K}_i\right)$	17, 21, 23

(17) The indices i and j label the molecular orbitals, and either μ or the numerals 1 and 2 label the electron coordinates.

(18) The double vertical bars denote an anti-symmetrized product of the occupied molecular spin orbitals $\phi_i\alpha$ and $\phi_i\beta$ (sometimes denoted ϕ_i and $\overline{\phi}_i$); for a closed-shell system Ψ would be a normalized Slater determinant. $(n!)^{-1/2}$ is the normalization factor and n the number of electrons.

(19) Z_A is the proton number (charge number) of nucleus A, and $r_{\mu A}$ is the distance of electron μ from nucleus A. H_{ii} is the energy of an electron in orbital ϕ_i in the field of the core.

(20) The inter-electron repulsion integral is written in various shorthand notations: In $J_{ij} = \langle ij|ij \rangle$ the first and third indices refer to the index of electron 1 and the second and fourth indices to electron 2. In $J_{ij} = (i^*i|j^*j)$, the first two indices refer to electron 1 and the second two indices to electron 2. Usually the functions are real and the stars are omitted. The exchange integral is written in various shorthand notations with the same index convention as described: $K_{ij} = \langle ij|ji \rangle$ or $K_{ij} = (i^*j|j^*i)$.

(21) These relations apply to closed-shell systems only, and the sums extend over the occupied molecular orbitals.

(22) The sum over j includes the term with $j = i$, for which $J_{ii} = K_{ii}$, so that this term in the sum simplifies to give $2J_{ii} - K_{ii} = J_{ii}$.

2.4.2 Hartree-Fock-Roothaan SCF theory, using molecular orbitals expanded as linear combinations of atomic-orbital basis functions (LCAO-MO theory)

Name	Symbol	Definition	Notes
atomic-orbital basis function	χ_r		24
molecular orbital	ϕ_i	$\phi_i = \sum_r \chi_r c_{ri}$	
overlap matrix element	S_{rs}	$S_{rs} = \int \chi_r^* \chi_s \mathrm{d}\tau, \quad \sum_{r,s} c_{ri}^* S_{rs} c_{sj} = \delta_{ij}$	
density matrix element	P_{rs}	$P_{rs} = 2 \sum_i^{\mathrm{occ}} c_{ri}^* c_{si}$	25
integrals over the basis functions:			
one-electron integrals	H_{rs}	$H_{rs} = \int \chi_r^*(1) \hat{H}_1^{\mathrm{core}} \chi_s(1) \mathrm{d}\tau_1$	
two-electron integrals	$(rs\|tu)$	$(rs\|tu) = \iint \chi_r^*(1)\,\chi_s(1)\,\frac{1}{r_{12}}\chi_t^*(2)\,\chi_u(2)\,\mathrm{d}\tau_1\mathrm{d}\tau_2$	26, 27
total electronic energy	E	$E = \sum_r \sum_s P_{rs} H_{rs}$ $+ (1/2) \sum_r \sum_s \sum_t \sum_u P_{rs} P_{tu}\left[(rs\|tu) - (1/2)(ru\|ts)\right]$	25, 27
matrix element of the Fock operator	F_{rs}	$F_{rs} = H_{rs} + \sum_t \sum_u P_{tu}\left[(rs\|tu) - (1/2)(ru\|ts)\right]$	25, 28

(Notes continued)

(23) The Hartree-Fock equations read $(\widehat{F} - \varepsilon_j)\phi_j = 0$. Note that the definition of the Fock operator involves all of its eigenfunctions ϕ_i through the coulomb and exchange operators, $\widehat{J}_i$ and $\widehat{K}_i$.

(24) The indices r and s label the basis functions. In numerical computations the basis functions are either taken as Slater-type orbitals (STO) or as Gaussian-type orbitals (GTO). An STO basis function in spherical polar coordinates has the general form $\chi(r, \theta, \phi) = N r^{n-1} \exp(-\zeta_{nl} r) Y_{lm}(\theta, \phi)$, where ζ_{nl} is a shielding parameter representing the effective charge in the state with quantum numbers n and l. GTO functions are typically expressed in cartesian space coordinates, in the form $\chi(x, y, z) = N x^a y^b z^c \exp(-\alpha r^2)$. Commonly, a linear combination of such functions with varying exponents α is used in such a way as to model an STO. N denotes a normalization factor.

(25) For closed-shell species with two electrons per occupied orbital. The sum extends over all occupied molecular orbitals. P_{rs} may also be called the bond order between atoms r and s.

(26) The contracted notation for two-electron integrals over the basis functions, $(rs|tu)$, is based on the same convention outlined in note 20.

(27) Here the two-electron integral is expressed in terms of integrals over the spatial atomic-orbital basis functions. The matrix elements H_{ii}, J_{ij}, and K_{ij} may be similarly expressed in terms of integrals over the spatial atomic-orbital basis functions according to the following equations:

$$H_{ii} = \sum_r \sum_s c_{ri}^* c_{si} H_{rs}$$
$$J_{ij} = (i^*i|j^*j) = \sum_r \sum_s \sum_t \sum_u c_{ri}^* c_{si} c_{tj}^* c_{uj} (r^*s|t^*u)$$
$$K_{ij} = (i^*j|j^*i) = \sum_r \sum_s \sum_t \sum_u c_{ri}^* c_{si} c_{tj}^* c_{uj} (r^*u|t^*s)$$

(28) The Hartree-Fock-Roothaan SCF equations, expressed in terms of the matrix elements of the Fock operator F_{rs}, and the overlap matrix elements S_{rs}, take the form:

$$\sum_s (F_{rs} - \varepsilon_i S_{rs})\, c_{si} = 0$$

2.5 ATOMS AND MOLECULES

The names and symbols recommended here are in agreement with those recommended by IUPAP [4] and ISO [5.i]. Additional quantities and symbols used in atomic, nuclear and plasma physics can be found in [4,5.j].

Name	Symbol	Definition	SI unit	Notes
nucleon number, mass number	A		1	
proton number, atomic number	Z		1	
neutron number	N	$N = A - Z$	1	
electron mass	m_e		kg	1, 2
mass of atom, atomic mass	m_a, m		kg	
atomic mass constant	m_u	$m_u = m_a(^{12}C)/12$	kg	1, 3
mass excess	Δ	$\Delta = m_a - Am_u$	kg	
elementary charge	e	proton charge	C	2
Planck constant	h		J s	
Planck constant divided by 2π	$\hbar$	$\hbar = h/2\pi$	J s	2
Bohr radius	a_0	$a_0 = 4\pi\varepsilon_0\hbar^2/m_e e^2$	m	2
Hartree energy	E_h	$E_h = \hbar^2/m_e a_0^2$	J	2
Rydberg constant	R_∞	$R_\infty = E_h/2hc$	m^{-1}	
fine-structure constant	α	$\alpha = e^2/4\pi\varepsilon_0\hbar c$	1	
ionization energy	E_i, I		J	4
electron affinity	E_{ea}, A		J	4
electronegativity	χ	$\chi = (1/2)(E_i + E_{ea})$	J	5
dissociation energy	E_d, D		J	
from the ground state	D_0		J	6
from the potential minimum	D_e		J	6

(1) Analogous symbols are used for other particles with subscripts: p for proton, n for neutron, a for atom, N for nucleus, etc.

(2) This quantity is also used as an atomic unit (see Section 3.9.1, p. 94 and Section 7.3 (iv), p. 145).

(3) m_u is equal to the unified atomic mass unit, with symbol u, i.e. $m_u = 1$ u (see Section 3.7, p. 92). The name dalton, with symbol Da, is used as an alternative name for the unified atomic mass unit [23].

(4) The ionization energy is frequently called the ionization potential (I_p). The electron affinity is the energy needed to detach an electron.

(5) The concept of electronegativity was introduced by L. Pauling as the power of an atom in a molecule to attract electrons to itself. There are several ways of defining this quantity [24]. The one given in the table has a clear physical meaning of energy and is due to R. S. Mulliken. The most frequently used scale, due to Pauling, is based on bond dissociation energies E_d in eV and it is relative in the sense that the values are dimensionless and that only electronegativity differences are defined. For atoms A and B

$$\chi_{r,A} - \chi_{r,B} = \sqrt{\frac{E_d(AB)}{eV} - \frac{1}{2}\frac{[E_d(AA) + E_d(BB)]}{eV}}$$

where χ_r denotes the Pauling relative electronegativity. The scale is chosen so as to make the relative electronegativity of hydrogen $\chi_{r,H} = 2.1$. There is a difficulty in choosing the sign of the square root, which determines the sign of $\chi_{r,A} - \chi_{r,B}$. Pauling made this choice intuitively.

(6) The symbols D_0 and D_e are used for dissociation energies of diatomic and polyatomic molecules.

Name	Symbol	Definition	SI unit	Notes
principal quantum number (hydrogen-like atom)	n	$E = hcZ^2 R_\infty/n^2$	1	7
angular momentum quantum numbers		see SPECTROSCOPY, Section 2.6		
magnetic dipole moment of a molecule	$\boldsymbol{m}, \boldsymbol{\mu}$	$E_p = -\boldsymbol{m} \cdot \boldsymbol{B}$	J T^{-1}	8
magnetizability of a molecule	ξ	$\boldsymbol{m} = \xi \boldsymbol{B}$	J T^{-2}	
Bohr magneton	μ_B	$\mu_B = e\hbar/2m_e$	J T^{-1}	
nuclear magneton	μ_N	$\mu_N = e\hbar/2m_p = (m_e/m_p)\,\mu_B$	J T^{-1}	
gyromagnetic ratio, (magnetogyric ratio)	γ	$\gamma_e = -g_e\mu_B/\hbar$	s^{-1} T^{-1}	9
g-factor	g, g_e	$g_e = -\gamma_e(2m_e/e)$	1	10
nuclear g-factor	g_N	$g_N = \gamma_N(2m_p/e)$	1	10
Larmor angular frequency	$\boldsymbol{\omega}_L$	$\boldsymbol{\omega}_L = -\gamma \boldsymbol{B}$	s^{-1}	11
Larmor frequency	ν_L	$\nu_L = \omega_L/2\pi$	Hz	
relaxation time,				
longitudinal	T_1		s	12
transverse	T_2		s	12
electric dipole moment of a molecule	$\boldsymbol{p}, \boldsymbol{\mu}$	$E_p = -\boldsymbol{p} \cdot \boldsymbol{E}$	C m	13
quadrupole moment of a molecule	$\boldsymbol{Q}; \boldsymbol{\Theta}$	$E_p = (1/2)\boldsymbol{Q} : \boldsymbol{V}'' = (1/3)\boldsymbol{\Theta} : \boldsymbol{V}''$	C m^2	14

(7) For an electron in the central coulomb field of an infinitely heavy nucleus of atomic number Z.

(8) Magnetic moments of specific particles may be denoted by subscripts, e.g. μ_e, μ_p, μ_n for an electron, a proton, and a neutron. Tabulated values usually refer to the maximum expectation value of the z component. Values for stable nuclei are given in Section 6.3, p. 121.

(9) The gyromagnetic ratio for a nucleus is $\gamma_N = g_N\mu_N/\hbar$.

(10) For historical reasons, $g_e > 0$. e is the (positive) elementary charge, therefore $\gamma_e < 0$. For nuclei, γ_N and g_N have the same sign. A different sign convention for the electronic g-factor is discussed in [25].

(11) This is a vector quantity with magnitude ω_L. This quantity is sometimes called Larmor circular frequency.

(12) These quantities are used in the context of saturation effects in spectroscopy, particularly spin-resonance spectroscopy (see Section 2.6, p. 27−28).

(13) See Section 2.6, note 9, p. 26.

(14) The quadrupole moment of a molecule may be represented either by the tensor $\boldsymbol{Q}$, defined by an integral over the charge density ρ:

$$Q_{\alpha\beta} = \int r_\alpha r_\beta \rho \; \mathrm{d}V$$

in which α and β denote x, y or z, or by the tensor $\boldsymbol{\Theta}$ of trace zero defined by

$$\Theta_{\alpha\beta} = (1/2) \int \left(3r_\alpha r_\beta - \delta_{\alpha\beta}\, r^2\right) \rho \; \mathrm{d}V = (1/2)\left[3Q_{\alpha\beta} - \delta_{\alpha\beta}\left(Q_{xx} + Q_{yy} + Q_{zz}\right)\right]$$

$\boldsymbol{V}''$ is the second derivative of the electronic potential:

$$V_{\alpha\beta}'' = -q_{\alpha\beta} = \partial^2 V/\partial\alpha\partial\beta$$

The contribution to the potential energy is then given by

$$E_p = (1/2)\boldsymbol{Q} : \boldsymbol{V}'' = (1/2)\sum_\alpha \sum_\beta Q_{\alpha\beta} V_{\alpha\beta}''$$

Name	Symbol	Definition	SI unit	Notes
quadrupole moment of a nucleus	eQ	$eQ = 2\langle\Theta_{zz}\rangle$	$C\ m^2$	15
electric field gradient tensor	$\mathbf{q}$	$q_{\alpha\beta} = -\partial^2 V/\partial\alpha\partial\beta$	$V\ m^{-2}$	
quadrupole interaction energy tensor	$\boldsymbol{\chi}$	$\chi_{\alpha\beta} = eQq_{\alpha\beta}$	J	16
electric polarizability of a molecule	$\boldsymbol{\alpha}$	$\alpha_{ab} = \partial p_a/\partial E_b$	$C^2\ m^2\ J^{-1}$	17
1st hyper-polarizability	$\boldsymbol{\beta}$	$\beta_{abc} = \partial^2 p_a/\partial E_b\partial E_c$	$C^3\ m^3\ J^{-2}$	17
2nd hyper-polarizability	$\boldsymbol{\gamma}$	$\gamma_{abcd} = \partial^3 p_a/\partial E_b\partial E_c\partial E_d$	$C^4\ m^4\ J^{-3}$	17
activity (of a radio-active substance)	A	$A = -dN_B/dt$	Bq	18
decay (rate) constant, disintegration (rate) constant	λ, k	$A = \lambda N_B$	s^{-1}	18
half life	$t_{1/2}, T_{1/2}$	$N_B(t_{1/2}) = N_B(0)/2$	s	18, 19
mean life, lifetime	τ	$\tau = 1/\lambda$	s	19
level width	Γ	$\Gamma = \hbar/\tau$	J	
disintegration energy	Q		J	
cross section	σ		m^2	
electroweak charge of a nucleus	Q_W	$Q_W \approx Z(1 - 4\sin^2\theta_W) - N$	1	20

(15) Nuclear quadrupole moments are conventionally defined in a different way from molecular quadrupole moments. Q has the dimension of an area and e is the elementary charge. eQ is taken to be twice the maximum expectation value of the zz tensor element (see note 14). The values of Q for some nuclei are listed in Section 6.3, p. 121.

(16) The nuclear quadrupole interaction energy tensor $\boldsymbol{\chi}$ is usually quoted in MHz, corresponding to the value of eQq/h, although the h is usually omitted.

(17) The polarizability $\boldsymbol{\alpha}$ and the hyper-polarizabilities $\boldsymbol{\beta}, \boldsymbol{\gamma}, \cdots$ are the coefficients in the expansion of the dipole moment $\mathbf{p}$ in powers of the electric field strength $\mathbf{E}$ (see Section 2.3, note 7, p. 16). The expansion of the component a is given by

$$p_a = p_a^{(0)} + \sum_b \alpha_{ab}E_b + (1/2)\sum_{bc}\beta_{abc}E_bE_c + (1/6)\sum_{bcd}\gamma_{abcd}E_bE_cE_d + \cdots$$

in which α_{ab}, β_{abc}, and γ_{abcd} are elements of the tensors $\boldsymbol{\alpha}$, $\boldsymbol{\beta}$, and $\boldsymbol{\gamma}$ of rank 2, 3, and 4, respectively. The components of these tensors are distinguished by the subscript indices $abc\cdots$, as indicated in the definitions, the first index a always denoting the component of p, and the subsequent indices the components of the electric field. The polarizability and the hyper-polarizabilities exhibit symmetry properties. Thus $\boldsymbol{\alpha}$ is commonly a symmetric tensor, and all components of $\boldsymbol{\beta}$ are zero for a molecule with a centre of symmetry, etc. Values of the polarizability are commonly quoted as the value $\alpha/4\pi\varepsilon_0$, which is a volume. The value is commonly expressed in the unit $Å^3$ (Å should not be used, see Section 2.6, note 11, p. 27) or in the unit a_0^3 (atomic units, see Section 3.9.1, p. 94). Similar comments apply to the hyper-polarizabilities with $\beta/(4\pi\varepsilon_0)^2$ in units of $a_0^5 e^{-1}$, and $\gamma/(4\pi\varepsilon_0)^3$ in units of $a_0^7 e^{-2}$, etc.

(18) N_B is the number of decaying entities B (1 Bq = 1 s^{-1}, see Section 3.4, p. 89).

(19) Half lives and mean lives are commonly given in years (unit a), see Section 7.2, note 4, p. 137. $t_{1/2} = \tau \ln 2$ for exponential decays.

(20) The electroweak charge of a nucleus is approximately given by the neutron number N and the proton number Z with the weak mixing angle θ_W (see Chapter 5, p. 111). It is important in calculations of atomic and molecular properties including the weak nuclear interaction [26].

2.6 SPECTROSCOPY

This section has been considerably extended compared with the original Manual [1.a–1.c] and with the corresponding section in the IUPAP document [4]. It is based on the recommendations of the ICSU Joint Commission for Spectroscopy [27, 28] and current practice in the field which is well represented in the books by Herzberg [29–31]. The IUPAC Commission on Molecular Structure and Spectroscopy has also published recommendations which have been taken into account [32–43].

Name	Symbol	Definition	SI unit	Notes
total term	T	$T = E_{\text{tot}}/hc$	m^{-1}	1, 2
transition wavenumber	$\widetilde{\nu}$	$\widetilde{\nu} = T' - T''$	m^{-1}	1
transition frequency	ν	$\nu = \left(E' - E''\right)/h$	Hz	
electronic term	T_e	$T_e = E_e/hc$	m^{-1}	1, 2
vibrational term	G	$G = E_{\text{vib}}/hc$	m^{-1}	1, 2
rotational term	F	$F = E_{\text{rot}}/hc$	m^{-1}	1, 2
spin-orbit coupling constant	A	$T_{\text{so}} = A <\widehat{\boldsymbol{L}} \cdot \widehat{\boldsymbol{S}}>$	m^{-1}	1, 3
principal moments of inertia	$I_A; I_B; I_C$	$I_A \leqslant I_B \leqslant I_C$	kg m^2	
rotational constants,				
in wavenumber	$\widetilde{A}; \widetilde{B}; \widetilde{C}$	$\widetilde{A} = h/8\pi^2 c I_A$	m^{-1}	1, 2
in frequency	$A; B; C$	$A = h/8\pi^2 I_A$	Hz	
inertial defect	Δ	$\Delta = I_C - I_A - I_B$	kg m^2	
asymmetry parameter	κ	$\kappa = \dfrac{2B - A - C}{A - C}$	1	4
centrifugal distortion constants,				
S reduction	$D_J; D_{JK}; D_K; d_1; d_2$		m^{-1}	5
A reduction	$\Delta_J; \Delta_{JK}; \Delta_K; \delta_J; \delta_K$		m^{-1}	5
harmonic vibration wavenumber	$\omega_e; \omega_r$		m^{-1}	6
vibrational anharmonicity constant	$\omega_e x_e; x_{rs}; g_{tt'}$		m^{-1}	6

(1) In spectroscopy the unit cm^{-1} is almost always used for the quantity wavenumber, and term values and wavenumbers always refer to the reciprocal wavelength of the equivalent radiation in vacuum. The symbol c in the definition E/hc refers to the speed of light in vacuum. Because "wavenumber" is not a number ISO suggests the use of repetency in parallel with wavenumber [5,6]. The use of the word "wavenumber" in place of the unit cm^{-1} must be avoided.

(2) Term values and rotational constants are sometimes defined in wavenumber units (e.g. $T = E/hc$), and sometimes in frequency units (e.g. $T = E/h$). When the symbol is otherwise the same, it is convenient to distinguish wavenumber quantities with a tilde (e.g. $\widetilde{\nu}, \widetilde{T}, \widetilde{A}, \widetilde{B}, \widetilde{C}$ for quantities defined in wavenumber units), although this is not a universal practice.

(3) $\widehat{\boldsymbol{L}}$ and $\widehat{\boldsymbol{S}}$ are electron orbital and electron spin operators, respectively.

(4) The Wang asymmetry parameters are also used: for a near prolate top $b_{\text{p}} = (C - B)/(2A - B - C)$, and for a near oblate top $b_{\text{o}} = (A - B)/(2C - A - B)$.

(5) S and A stand for the symmetric and asymmetric reductions of the rotational hamiltonian respectively; see [44] for more details on the various possible representations of the centrifugal distortion constants.

(6) For a diatomic molecule: $G(v) = \omega_e \left(v + \frac{1}{2}\right) - \omega_e x_e \left(v + \frac{1}{2}\right)^2 + \cdots$. For a polyatomic molecule the $3N - 6$ vibrational modes ($3N - 5$ if linear) are labeled by the indices $r, s, t, \cdots$, or $i, j, k, \cdots$. The index r is usually assigned to be increasing with descending wavenumber, symmetry species by

Name	Symbol	Definition	SI unit	Notes
vibrational quantum numbers	$v_r; l_t$		1	6
vibrational fundamental wavenumber	$\widetilde{\nu}_r, \widetilde{\nu}_r^0, \nu_r$	$\widetilde{\nu}_r = T(v_r = 1) - T(v_r = 0)$	m^{-1}	6
Coriolis ζ-constant	ζ_{rs}^α		1	7
angular momentum quantum numbers		see Section 2.6.1, p. 30		
degeneracy, statistical weight	g, d, β		1	8
electric dipole moment of a molecule	$\boldsymbol{p}, \boldsymbol{\mu}$	$E_p = -\boldsymbol{p} \cdot \boldsymbol{E}$	C m	9
transition dipole moment of a molecule	$\boldsymbol{M}, \boldsymbol{R}$	$\boldsymbol{M} = \int \psi^{*\prime} \boldsymbol{p} \, \psi^{\prime\prime} \, \mathrm{d}\tau$	C m	9, 10

(6) (continued) symmetry species starting with the totally symmetric species. The index t is kept for degenerate modes. The vibrational term formula is

$$G\left(v\right) = \sum_r \omega_r \left(v_r + d_r/2\right) + \sum_{r \leqslant s} x_{rs} \left(v_r + d_r/2\right) \left(v_s + d_s/2\right) + \sum_{t \leqslant t'} g_{tt'} l_t l_{t'} + \cdots$$

Another common term formula is defined with respect to the vibrational ground state

$$T(v) = \sum_i v_i \widetilde{\nu}_i{}' + \sum_{r \leqslant s} x_{rs}{}' v_r v_s + \sum_{t \leqslant t'} g_{tt'}{}' l_t l_{t'} + \cdots$$

these being commonly used as the diagonal elements of effective vibrational hamiltonians.

(7) Frequently the Coriolis coupling constants $\xi_\alpha^{v'v}$ with units of cm^{-1} are used as effective hamiltonian constants ($\alpha = x, y, z$). For two fundamental vibrations with harmonic wavenumbers ω_r and ω_s these are connected with ζ_{rs}^α by the equation ($v_r = 1$ and $v_s = 1$)

$$\xi_\alpha^{v'v} = \widetilde{B}_\alpha \, \zeta_{rs}^\alpha \left[\sqrt{\omega_s/\omega_r} + \sqrt{\omega_r/\omega_s}\right]$$

where $\widetilde{B}_\alpha$ is the α rotational constant. A similar equation applies with B_α if $\xi_\alpha^{v'v}$ is defined as a quantity with frequency units.

(8) d is usually used for vibrational degeneracy, and β for nuclear spin degeneracy.

(9) E_p is the potential energy of a dipole with electric dipole moment $\boldsymbol{p}$ in an electric field of strength $\boldsymbol{E}$. Molecular dipole moments are commonly expressed in the non-SI unit debye, where $1 \ D \approx 3.335 \ 64 \times 10^{-30}$ C m. The SI unit C m is inconvenient for expressing molecular dipole moments, which results in the continued use of the deprecated debye (D). A convenient alternative is to use the atomic unit, ea_0. Another way of expressing dipole moments is to quote the electric dipole lengths, $l_p = p/e$, analogous to the way the nuclear quadrupole areas are quoted (see Section 2.5, notes 14 and 15, p. 23 and 24 and Section 6.3, p. 121). This gives the distance between two elementary charges of the equivalent dipole and conveys a clear picture in relation to molecular dimensions (see also Section 2.3, note 8, p. 17).

	Dipole moment			Dipole length
Examples	SI		a.u.	
	p/C m	p/D	p/ea_0	l_p/pm
HCl	3.60×10^{-30}	1.08	0.425	22.5
H_2O	6.23×10^{-30}	1.87	0.736	38.9
NaCl	3.00×10^{-29}	9.00	3.54	187

(10) For quantities describing line and band intensities see Section 2.7, p. 34−41.

Name	Symbol	Definition	SI unit	Notes
interatomic distances,				11, 12
equilibrium	r_e		m	
zero-point average	r_z		m	
ground state	r_0		m	
substitution structure	r_s		m	
vibrational coordinates,				11
internal	R_i, r_i, θ_j, etc.		(varies)	
symmetry	S_j		(varies)	
normal				
mass adjusted	Q_r		$\mathrm{kg}^{1/2}$ m	
dimensionless	q_r		1	
vibrational force constants,				
diatomic	$f, (k)$	$f = \partial^2 V/\partial r^2$	J m^{-2}, N m^{-1}	11, 13
polyatomic,				
internal coordinates	f_{ij}	$f_{ij} = \partial^2 V/\partial r_i \partial r_j$	(varies)	
symmetry coordinates	F_{ij}	$F_{ij} = \partial^2 V/\partial S_i \partial S_j$	(varies)	
dimensionless normal	$\phi_{rst\cdots}, k_{rst\cdots},$		m^{-1}	14
coordinates	$C_{rst\cdots}$			

electron spin resonance (ESR), electron paramagnetic resonance (EPR):

Name	Symbol	Definition	SI unit	Notes
gyromagnetic ratio	γ	$\gamma = \mu/\hbar\sqrt{S(S+1)}$	s^{-1} T^{-1}	15
g-factor	g	$h\nu = g\mu_B B$	1	16
hyperfine coupling constant				
in liquids	a, A	$\widehat{H}_{\mathrm{hfs}}/h = a\,\widehat{\boldsymbol{S}}\cdot\widehat{\boldsymbol{I}}$	Hz	17
in solids	$\boldsymbol{T}$	$\widehat{H}_{\mathrm{hfs}}/h = \widehat{\boldsymbol{S}}\cdot\boldsymbol{T}\cdot\widehat{\boldsymbol{I}}$	Hz	17

(11) Interatomic (internuclear) distances and vibrational displacements are commonly expressed in the non-SI unit ångström, where 1 Å = 10^{-10} m = 0.1 nm = 100 pm. Å should not be used.

(12) The various slightly different ways of representing interatomic distances, distinguished by subscripts, involve different vibrational averaging contributions; they are discussed in [45], where the geometrical structures of many free molecules are listed. Only the equilibrium distance r_e is isotopically invariant, to a good approximation. The effective distance parameter r_0 is estimated from the rotational constants for the ground vibrational state and has a more complicated physical significance for polyatomic molecules.

(13) Force constants are commonly expressed in mdyn Å^{-1} = aJ Å^{-2} for stretching coordinates, mdyn Å = aJ for bending coordinates, and mdyn = aJ Å^{-1} for stretch-bend interactions. Å should not be used (see note 11). See [46] for further details on definitions and notation for force constants.

(14) The force constants in dimensionless normal coordinates are usually defined in wavenumber units by the equation $V/hc = \sum \phi_{rst\cdots} q_r q_s q_t \cdots$, where the summation over the normal coordinate indices $r, s, t, \cdots$ is unrestricted.

(15) The magnitude of γ is obtained from the magnitudes of the magnetic dipole moment μ and the angular momentum (in ESR: electron spin $\boldsymbol{S}$ for a Σ-state, $L = 0$; in NMR: nuclear spin $\boldsymbol{I}$).

(16) This gives an effective g-factor for a single spin ($S = 1/2$) in an external static field (see Section 2.5, note 10, p. 23).

(17) $\widehat{H}_{\mathrm{hfs}}$ is the hyperfine coupling hamiltonian. $\widehat{\boldsymbol{S}}$ is the electron spin operator with quantum number S (see Section 2.6.1, p. 30). The coupling constants a are usually quoted in MHz, but they are sometimes quoted in magnetic induction units (G or T) obtained by dividing by the conversion factor $g\mu_B/h$, which has the SI unit Hz/T; $g_e\mu_B/h \approx 28.025$ GHz T^{-1} ($= 2.8025$ MHz G^{-1}), where g_e is the g-factor for a free electron. If in liquids the hyperfine coupling is isotropic, the coupling

Name	Symbol	Definition	SI unit	Notes
nuclear magnetic resonance (NMR):				
static magnetic flux density of a NMR spectrometer	$\boldsymbol{B}_0$		T	18
radiofrequency magnetic flux densities	$\boldsymbol{B}_1, \boldsymbol{B}_2$		T	18
spin-rotation interaction tensor	$\boldsymbol{C}$		Hz	19
spin-rotation coupling constant of nucleus A	C_A		Hz	19
dipolar interaction tensor	$\boldsymbol{D}$		Hz	19
dipolar coupling constant between nuclei A, B	D_{AB}	$D_{AB} = \dfrac{\mu_0 \hbar}{8\pi^2} \dfrac{\gamma_A \gamma_B}{r_{AB}^3}$	Hz	19
frequency variables of a two-dimensional spectrum	$F_1, F_2; f_1, f_2$		Hz	
nuclear spin operator for nucleus A	$\widehat{\boldsymbol{I}}_A, \widehat{\boldsymbol{S}}_A$		1	20
quantum number associated with $\widehat{\boldsymbol{I}}_A$	I_A		1	
indirect spin coupling tensor	$\boldsymbol{J}$		Hz	
nuclear spin-spin coupling through n bonds	^nJ		Hz	21
reduced nuclear spin-spin coupling constant	K_{AB}	$K_{AB} = \dfrac{J_{AB}}{h} \dfrac{2\pi}{\gamma_A} \dfrac{2\pi}{\gamma_B}$	$T^2\ J^{-1}$, N A^{-2} m^{-3}	20
eigenvalue of $\widehat{\boldsymbol{I}}_{Az}$	m_A		1	22
equilibrium macroscopic magnetization per volume	$\boldsymbol{M}_0$		J T^{-1} m^{-3}	23
nuclear quadrupole moment	eQ		C m^2	24

(17) (continued) constant is a scalar a. In solids the coupling is anisotropic, and the coupling constant is a 3×3 tensor $\boldsymbol{T}$. Similar comments apply to the g-factor and to the analogous NMR parameters. A convention for the g-factor has been proposed, in which the sign of g is positive when the dipole moment is parallel to its angular momentum and negative, when it is anti parallel. Such a choice would require the g-factors for the electron orbital and spin angular momenta to be negative [25] (see Section 2.5, note 10, p. 23).

(18) The observing ($\boldsymbol{B}_1$) and irradiating ($\boldsymbol{B}_2$) radiofrequency flux densities are associated with frequencies ν_1, ν_2 and with nutation angular frequencies Ω_1, Ω_2 (around $\boldsymbol{B}_1$, $\boldsymbol{B}_2$, respectively). They are defined through $\Omega_1 = -\gamma \boldsymbol{B}_1$ and $\Omega_2 = -\gamma \boldsymbol{B}_2$ (see Section 2.3, notes 8 and 9, p. 17).

(19) The units of interaction strengths are Hz. In the context of relaxation the interaction strengths should be converted to angular frequency units (rad s^{-1}, but commonly denoted s^{-1}, see note 25).

(20) $\widehat{\boldsymbol{I}}_A$, $\widehat{\boldsymbol{S}}_A$ have components $\widehat{I}_{Ax}$, $\widehat{I}_{Ay}$, $\widehat{I}_{Az}$ or $\widehat{S}_{Ax}$, $\widehat{S}_{Ay}$, $\widehat{S}_{Az}$, respectively. For another way of defining angular momentum operators with unit J s (see Section 2.6.1, p. 30).

(21) Parentheses may be used to indicate the species of the nuclei coupled, e.g. $J(^{13}C, {}^1H)$ or, additionally, the coupling path, e.g. $J(POCF)$. Where no ambiguity arises, the elements involved can be, alternatively, given as subscripts, e.g. J_{CH}. The nucleus of higher mass should be given first. The same applies to the reduced coupling constant.

(22) M rather than m is frequently recommended, but most NMR practitioners use m so as to avoid confusion with magnetization.

(23) This is for a spin system in the presence of a magnetic flux density $\boldsymbol{B}_0$.

(24) See Section 2.5, notes 14 and 15, p. 23 and 24.

Name	Symbol	Definition	SI unit	Notes
electric field gradient	$\boldsymbol{q}$	$q_{\alpha\beta} = -\partial^2 V/\partial\alpha\partial\beta$	V m^{-2}	25
nuclear quadrupole coupling constant	χ	$\chi = eq_{zz}Q/h$	Hz	19
time dimensions for two-dimensional NMR	t_1, t_2		s	
spin-lattice relaxation time (longitudinal) for nucleus A	$T_1{}^{\mathrm{A}}$		s	26, 27
spin-spin relaxation time (transverse) for nucleus A	$T_2{}^{\mathrm{A}}$		s	26
spin wavefunctions	$\alpha; \beta$			28
gyromagnetic ratio	γ	$\gamma = \mu/\hbar\sqrt{I(I+1)}$	s^{-1} T^{-1}	15, 29
chemical shift for the nucleus A	δ_{A}	$\delta_{\mathrm{A}} = (\nu_{\mathrm{A}} - \nu_{\mathrm{ref}})/\nu_{\mathrm{ref}}$	1	30
nuclear Overhauser enhancement	η		1	31
nuclear magneton	μ_{N}	$\mu_{\mathrm{N}} = (m_{\mathrm{e}}/m_{\mathrm{p}})\mu_{\mathrm{B}}$	J T^{-1}	32
resonance frequency	ν		Hz	18
of a reference molecule	ν_{ref}			
of the observed rf field	ν_1			
of the irradiating rf field	ν_2			
standardized resonance frequency for nucleus A	Ξ_{A}		Hz	33
shielding tensor	$\boldsymbol{\sigma}$		1	34
shielding constant	σ_{A}	$B_{\mathrm{A}} = (1 - \sigma_{\mathrm{A}})B_0$	1	34
correlation time	τ_{c}		s	26

(25) The symbol $\boldsymbol{q}$ is recommended by IUPAC for the field gradient tensor, with the units of V m^{-2} (see Section 2.5, p. 24). With $\boldsymbol{q}$ defined in this way, the quadrupole coupling constant is $\chi = eq_{zz}Q/h$. It is common in NMR to denote the field gradient tensor as eq and the quadrupole coupling constant as $\chi = e^2 q_{zz}Q/h$. $\boldsymbol{q}$ has principal components q_{XX}, q_{YY}, q_{ZZ}.

(26) The relaxation times and the correlation times are normally given in the units of s. Strictly speaking, this refers to s rad^{-1}.

(27) The spin-lattice relaxation time of nucleus A in the frame of reference rotating with $\boldsymbol{B}_1$ is denoted $T_{1\rho}{}^{\mathrm{A}}$.

(28) See Section 2.4, note 9, p. 19.

(29) μ is the magnitude of the nuclear magnetic moment.

(30) Chemical shift (of the resonance) for the nucleus of element A (positive when the sample resonates to high frequency of the reference); for conventions, see [47]. Further information regarding solvent, references, or nucleus of interest may be given by superscripts or subscripts or in parentheses.

(31) The nuclear Overhauser effect, $1 + \eta$, is defined as the ratio of the signal intensity for the I nuclear spin, obtained under conditions of saturation of the S nuclear spin, and the equilibrium signal intensity for the I nuclear spin.

(32) m_{e} and m_{p} are the electron and proton mass, respectively. μ_{B} is the Bohr magneton (see Section 2.5, notes 1 and 2, p. 22 and Chapter 5, p. 111).

(33) Resonance frequency for the nucleus of element A in a magnetic field such that the protons in tetramethylsilane (TMS) resonate exactly at 100 MHz.

(34) The symbols σ_{A} (and related terms of the shielding tensor and its components) should refer to shielding on an absolute scale (for theoretical work). For shielding relative to a reference, symbols such as $\sigma_{\mathrm{A}} - \sigma_{\mathrm{ref}}$ should be used. B_{A} is the corresponding effective magnetic flux density (see Section 2.3, note 9, p. 17).

2.6.1 Symbols for angular momentum operators and quantum numbers

In the following table, all of the operator symbols denote the dimensionless ratio *angular momentum* divided by $\hbar$. Although this is a universal practice for the quantum numbers, some authors use the operator symbols to denote *angular momentum*, in which case the operators would have SI units: J s. The column heading "*Z-axis*" denotes the space-fixed component, and the heading "*z-axis*" denotes the molecule-fixed component along the symmetry axis (linear or symmetric top molecules), or the axis of quantization.

Angular momentum[1]	Operator symbol	Quantum number symbol			Notes
		Total	Z-axis	z-axis	
electron orbital	$\widehat{\boldsymbol{L}}$	L	M_L	Λ	2
one electron only	$\widehat{\boldsymbol{l}}$	l	m_l	λ	2
electron spin	$\widehat{\boldsymbol{S}}$	S	M_S	Σ	
one electron only	$\widehat{\boldsymbol{s}}$	s	m_s	σ	
electron orbital plus spin	$\widehat{\boldsymbol{L}}+\widehat{\boldsymbol{S}}$			$\Omega = \Lambda + \Sigma$	2
nuclear orbital (rotational)	$\widehat{\boldsymbol{R}}$	R		K_R, k_R	
nuclear spin	$\widehat{\boldsymbol{I}}$	I	M_I		
internal vibrational					
spherical top	$\widehat{\boldsymbol{l}}$	$l\,(l\zeta)$		K_l	3
other	$\widehat{\boldsymbol{j}}, \widehat{\boldsymbol{\pi}}$			$l\,(l\zeta)$	2, 3
sum of $R + L\,(+j)$	$\widehat{\boldsymbol{N}}$	N		K, k	2
sum of $N + S$	$\widehat{\boldsymbol{J}}$	J	M_J	K, k	2, 4
sum of $J + I$	$\widehat{\boldsymbol{F}}$	F	M_F		

(1) In all cases the vector operator and its components are related to the quantum numbers by eigenvalue equation analogous to:

$$\widehat{\boldsymbol{J}}^2\psi = J\,(J+1)\,\psi, \quad \widehat{J}_Z\psi = M_J\psi, \quad \text{and} \quad \widehat{J}_z\,\psi = K\psi,$$

where the component quantum numbers M_J and K take integral or half-odd values in the range $-J \leqslant M_J \leqslant +J, -J \leqslant K \leqslant +J$. (If the operator symbols are taken to represent *angular momentum*, rather than *angular momentum* divided by $\hbar$, the eigenvalue equations should read $\widehat{\boldsymbol{J}}^2\psi = J\,(J+1)\,\hbar^2\psi$, $\widehat{J}_Z\psi = M_J\hbar\psi$, and $\widehat{J}_z\psi = K\hbar\psi$.) l is frequently called the azimuthal quantum number and m_l the magnetic quantum number.

(2) Some authors, notably Herzberg [29–31], treat the component quantum numbers Λ, Ω, l and K as taking positive or zero values only, so that each non-zero value of the quantum number labels two wavefunctions with opposite signs for the appropriate angular momentum component. When this is done, lower case k is commonly regarded as a signed quantum number, related to K by $K = |k|$. However, in theoretical discussions all component quantum numbers are usually treated as signed, taking both positive and negative values.

(3) There is no uniform convention for denoting the internal vibrational angular momentum; j, π, p and G have all been used. For symmetric top and linear molecules the component of j in the symmetry axis is always denoted by the quantum number l, where l takes values in the range $-v \leqslant l \leqslant +v$ in steps of 2. The corresponding component of angular momentum is actually $l\zeta\hbar$, rather than $l\hbar$, where ζ is the Coriolis ζ-constant (see note 7, p. 26).

(4) Asymmetric top rotational states are labeled by the value of J (or N if $S \neq 0$), with subscripts K_a, K_c, where the latter correlate with the $K = |k|$ quantum number about the a and c axes in the prolate and oblate symmetric top limits respectively.

> *Example* $J_{K_a,K_c} = 5_{2,3}$ for a particular rotational level.

2.6.2 Symbols for symmetry operators and labels for symmetry species

(i) Symmetry operators in space-fixed coordinates [41, 48]

identity	E
permutation	P, p
space-fixed inversion	$E^*, (P)$
permutation-inversion	$P^* (= PE^*), p^*$

The permutation operation P permutes the labels of identical nuclei.

Example In the NH_3 molecule, if the hydrogen nuclei are labeled 1, 2 and 3, then $P = (123)$ would symbolize the permutation where 1 is replaced by 2, 2 by 3, and 3 by 1.

The inversion operation E^* reverses the sign of all particle coordinates in the space-fixed origin, or in the molecule-fixed centre of mass if translation has been separated. It is also called the parity operator and then frequently denoted by P, although this cannot be done in parallel with P for permutation, which then should be denoted by lower case p. In field-free space and in the absence of parity violation [26], true eigenfunctions of the hamiltonian are either of positive parity $+$ (unchanged) or of negative parity $-$ (change sign) under E^*. The label may be used to distinguish the two nearly degenerate components formed by Λ-doubling (in a degenerate electronic state) or l-doubling (in a degenerate vibrational state) in linear molecules, or by K-doubling (asymmetry-doubling) in slightly asymmetric tops. For linear molecules, Λ- or l-doubled components may also be distinguished by the labels e or f [49]; for singlet states these correspond respectively to parity $+$ or $-$ for J even and vice versa for J odd (but see [49]). For linear molecules in degenerate electronic states the Λ-doubled levels may alternatively be labeled $\Pi (A')$ or $\Pi (A'')$ (or $\Delta (A')$, $\Delta (A'')$ etc.) [50]. Here the labels A' or A'' describe the symmetry of the electronic wavefunction at high J with respect to reflection in the plane of rotation (but see [50] for further details). The A' or A'' labels are particularly useful for the correlation of states of molecules involved in reactions or photodissociation.

In relation to permutation-inversion symmetry species the superscript $+$ or $-$ may be used to designate parity, whereas a letter is used to designate symmetry with respect to the permutation group. One can also use the systematic notation from the theory of the symmetric group (permutation group) S_n [51], the permutation inversion group being denoted by $S_n{}^*$ in this case, if one considers the full permutation group. The species is then given by the partition $P(S_n)$ [52–54]. The examples give the species for $S_4{}^*$, where the partition is conventionally given in square brackets [ ]. Conventions with respect to these symbols still vary ([2.b] and [39–41, 51–54]).

Examples A_1^+ totally symmetric species with respect to permutation, positive parity, $[4]^+$
A_1^- totally symmetric species with respect to permutation, negative parity, $[4]^-$
E^+ doubly degenerate species with respect to permutation, positive parity, $[2^2]^+$
E^- doubly degenerate species with respect to permutation, negative parity, $[2^2]^-$
F_1^+ triply degenerate species with respect to permutation, positive parity, $[2,1^2]^+$
F_1^- triply degenerate species with respect to permutation, negative parity, $[2,1^2]^-$

The Hermann-Mauguin symbols of symmetry operations used for crystals are given in Section 2.8.1 (ii), p. 44.

(ii) Symmetry operators in molecule-fixed coordinates (Schönflies symbols) [29–31]

identity	E
rotation by $2\pi/n$	C_n
reflection	$\sigma, \sigma_v, \sigma_d, \sigma_h$
inversion	i
rotation-reflection	$S_n (= C_n \sigma_h)$

If C_n is the primary axis of symmetry, wavefunctions that are unchanged or change sign under the operator C_n are given species labels A or B respectively, and otherwise wavefunctions that are multiplied by $\exp(\pm 2\pi i s/n)$ are given the species label E_s. Wavefunctions that are unchanged or change sign under i are labeled g (gerade) or u (ungerade) respectively. Wavefunctions that are unchanged or change sign under σ_h have species labels with a prime $'$ or a double prime $''$, respectively. For more detailed rules see [28–31].

2.6.3 Other symbols and conventions in optical spectroscopy

(i) Term symbols for atomic states

The electronic states of atoms are labeled by the value of the quantum number L for the state. The value of L is indicated by an upright capital letter: S, P, D, F, G, H, I, and K, $\cdots$, are used for $L = 0$, 1, 2, 3, 4, 5, 6, and 7, $\cdots$, respectively. The corresponding lower case letters are used for the orbital angular momentum of a single electron. For a many-electron atom, the electron spin multiplicity $(2S + 1)$ may be indicated as a left-hand superscript to the letter, and the value of the total angular momentum J as a right-hand subscript. If either L or S is zero only one value of J is possible, and the subscript is then usually suppressed. Finally, the electron configuration of an atom is indicated by giving the occupation of each one-electron orbital as in the examples below.

> *Examples*   B: $(1s)^2 (2s)^2 (2p)^1$, $^2P^\circ_{1/2}$
> C: $(1s)^2 (2s)^2 (2p)^2$, $^3P_0$
> N: $(1s)^2 (2s)^2 (2p)^3$, $^4S^\circ$

A right superscript $^\circ$ may be used to indicate odd parity (negative parity $-$). Omission of the superscript ^{e} is then to be interpreted as even parity (positive parity $+$). In order to avoid ambiguities it can be useful to always designate parity by $+$ or $-$ as a right superscript (i.e. $^4S^-_{3/2}$ for N and $^3P^+_0$ for C).

(ii) Term symbols for molecular states

The electronic states of molecules are labeled by the symmetry species label of the wavefunction in the molecular point group. These should be Latin or Greek upright capital letters. As for atoms, the spin multiplicity $(2S + 1)$ may be indicated by a left superscript. For linear molecules the value of $\Omega (= \Lambda + \Sigma)$ may be added as a right subscript (analogous to J for atoms). If the value of Ω is not specified, the term symbol is taken to refer to all component states, and a right subscript r or i may be added to indicate that the components are regular (energy increases with Ω) or inverted (energy decreases with Ω) respectively.

The electronic states of molecules are also given empirical single letter labels as follows. The ground electronic state is labeled X, excited states of the same multiplicity are labeled A, B, C, $\cdots$, in ascending order of energy, and excited states of different multiplicity are labeled with lower case letters a, b, c, $\cdots$. In polyatomic molecules (but not diatomic molecules) it is customary to add a tilde (e.g. $\tilde{X}$) to these empirical labels to prevent possible confusion with the symmetry species label.

Finally the one-electron orbitals are labeled by the corresponding lower case letters, and the electron configuration is indicated in a manner analogous to that for atoms.

> *Examples*   The ground state of CH is $(1\sigma)^2 (2\sigma)^2 (3\sigma)^2 (1\pi)^1$, X $^2\Pi_r$, in which the $^2\Pi_{1/2}$
> component lies below the $^2\Pi_{3/2}$ component, as indicated by the subscript r
> for regular.

Examples (continued)

The ground state of OH is $(1\sigma)^2 (2\sigma)^2 (3\sigma)^2 (1\pi)^3$, X $^2\Pi_i$, in which the $^2\Pi_{3/2}$ component lies below the $^2\Pi_{1/2}$ component, as indicated by the subscript i for inverted.

The two lowest electronic states of CH_2 are $\cdots (2a_1)^2 (1b_2)^2 (3a_1)^2$, $\tilde{a}\ ^1A_1$, $\cdots (2a_1)^2 (1b_2)^2 (3a_1)^1 (1b_1)^1$, $\tilde{X}\ ^3B_1$.

The ground state of C_6H_6 (benzene) is $\cdots (a_{2u})^2 (e_{1g})^4$, $\tilde{X}\ ^1A_{1g}$.

The vibrational states of molecules are usually indicated by giving the vibrational quantum numbers for each normal mode.

Examples For a bent triatomic molecule

$(0,0,0)$ denotes the ground state,

$(1,0,0)$ denotes the ν_1 state, i.e. $v_1 = 1$, and

$(1,2,0)$ denotes the $\nu_1 + 2\nu_2$ state, i.e. $v_1 = 1, v_2 = 2$, etc.

(iii) Notation for spectroscopic transitions

The upper and lower levels of a spectroscopic transition are indicated by a prime $'$ and double prime $''$ respectively.

Example $h\nu = E' - E''$

Transitions are generally indicated by giving the excited-state label, followed by the ground-state label, separated by a dash or an arrow to indicate the direction of the transition (emission to the right, absorption to the left).

Examples

B−A	indicates a transition between a higher energy state B and a lower energy state A;
B→A	indicates emission from B to A;
B←A	indicates absorption from A to B;
A→B	indicates more generally a transition from the initial state A to final state B in kinetics (see Section 2.12, p. 63);
$(0,2,1) \leftarrow (0,0,1)$	labels the $2\nu_2 + \nu_3 - \nu_3$ hot band in a bent triatomic molecule.

A more compact notation [55] may be used to label vibronic (or vibrational) transitions in polyatomic molecules with many normal modes, in which each vibration index r is given a superscript v'_r and a subscript v''_r indicating the upper electronic and the lower electronic state values of the vibrational quantum number. When $v'_r = v''_r = 0$ the corresponding index is suppressed.

Examples 1^1_0 denotes the transition $(1,0,0)-(0,0,0)$;

$2^2_0\ 3^1_1$ denotes the transition $(0,2,1)-(0,0,1)$.

In order to denote transitions within the same electronic state one may use matrix notation or an arrow.

Example 2_{20} or $2_{2\leftarrow 0}$ denotes a vibrational transition within the electronic ground state from $v_2 = 0$ to $v_2 = 2$.

For rotational transitions, the value of $\Delta J = J' - J''$ is indicated by a letter labelling the branches of a rotational band: $\Delta J = -2, -1, 0, 1$, and 2 are labelled as the O-branch, P-branch, Q-branch, R-branch, and S-branch respectively. The changes in other quantum numbers (such as K for a symmetric top, or K_a and K_c for an asymmetric top) may be indicated by adding lower case letters as a left superscript according to the same rule.

Example ^PQ labels a "p-type Q-branch" in a symmetric top molecule, i.e. $\Delta K = -1, \Delta J = 0$.

The value of K in the lower level is indicated as a right subscript, e.g. $^PQ_{K''}$ or $^PQ_2(5)$ indicating the transition from $K'' = 2$ to $K' = 1$, the value of J'' being added in parentheses.

2.7 ELECTROMAGNETIC RADIATION

The quantities and symbols given here have been selected on the basis of recommendations by IUPAP [4], ISO [5.f], and IUPAC [56–59], as well as by taking into account the practice in the field of laser physics. Terms used in photochemistry [60] have been considered as well, but definitions still vary. Terms used in high energy ionizing electromagnetic radiation, radiation chemistry, radiochemistry, and nuclear chemistry are not included or discussed.

Name	Symbol	Definition	SI unit	Notes
wavelength	λ		m	
speed of light				
in vacuum	c_0	$c_0 = 299\ 792\ 458\ \text{m s}^{-1}$	m s^{-1}	1
in a medium	c	$c = c_0/n$	m s^{-1}	1
wavenumber in vacuum	$\tilde{\nu}$	$\tilde{\nu} = \nu/c_0 = 1/\lambda$	m^{-1}	2
wavenumber	σ	$\sigma = 1/\lambda$	m^{-1}	
(in a medium)				
frequency	ν	$\nu = c/\lambda$	Hz	
angular frequency,	ω	$\omega = 2\pi\nu$	s^{-1}, rad s^{-1}	
pulsatance				
refractive index	n	$n = c_0/c$	1	
Planck constant	h		J s	
Planck constant divided by 2π	$\hbar$	$\hbar = h/2\pi$	J s	
radiant energy	Q, W		J	3
radiant energy density	ρ, w	$\rho = \mathrm{d}Q/\mathrm{d}V$	J m^{-3}	3
spectral radiant energy density				3
in terms of frequency	ρ_ν, w_ν	$\rho_\nu = \mathrm{d}\rho/\mathrm{d}\nu$	J m^{-3} Hz^{-1}	
in terms of wavenumber	$\rho_{\tilde{\nu}}, w_{\tilde{\nu}}$	$\rho_{\tilde{\nu}} = \mathrm{d}\rho/\mathrm{d}\tilde{\nu}$	J m^{-2}	
in terms of wavelength	ρ_λ, w_λ	$\rho_\lambda = \mathrm{d}\rho/\mathrm{d}\lambda$	J m^{-4}	
radiant power (radiant energy	P, Φ	$P = \mathrm{d}Q/\mathrm{d}t$	W	3
per time)				3
radiant intensity	I_e	$I_e = \mathrm{d}P/\mathrm{d}\Omega$	W sr^{-1}	3, 4
radiant excitance	M	$M = \mathrm{d}P/\mathrm{d}A_{\text{source}}$	W m^{-2}	3, 4

(1) When there is no risk of ambiguity the subscript denoting vacuum is often omitted. n denotes the refraction index of the medium.

(2) The unit cm^{-1} is widely used for the quantity wavenumber in vacuum.

(3) The symbols for the quantities such as *radiant energy* and *intensity* are also used for the corresponding quantities concerning visible radiation, i.e. luminous quantities and photon quantities. Subscripts e for energetic, v for visible, and p for photon may be added whenever confusion between these quantities might otherwise occur. The units used for luminous quantities are derived from the base unit candela (cd) (see Section 3.3, p. 87).

Examples	radiant intensity	I_e, $I_e = \mathrm{d}P/\mathrm{d}\Omega$,	SI unit: W sr^{-1}
	luminous intensity	I_v,	SI unit: cd
	photon intensity	I_p,	SI unit: s^{-1} sr^{-1}

The radiant intensity I_e should be distinguished from the plain intensity or irradiance I (see note 5). Additional subscripts to distinguish absorbed (abs), transmitted (tr) or reflected (refl) quantities may be added, if necessary.

(4) The radiant intensity is the radiant power per solid angle in the direction of the point from which the source is being observed. The radiant excitance is the total emitted radiant power per area A_{source} of the radiation source, for all wavelengths. The radiance is the radiant intensity per area of

34

Name	Symbol	Definition	SI unit	Notes
radiance	L	$I_e = \int L \cos \Theta \, \mathrm{d}A_{\text{source}}$	W sr^{-1} m^{-2}	3, 4
intensity, irradiance	I, E	$I = \mathrm{d}P/\mathrm{d}A$	W m^{-2}	3, 5
spectral intensity, spectral irradiance	$I_{\tilde{\nu}}, E_{\tilde{\nu}}$	$I_{\tilde{\nu}} = \mathrm{d}I/\mathrm{d}\tilde{\nu}$	W m^{-1}	6
fluence	$F, (H)$	$F = \int I \mathrm{d}t = \int (\mathrm{d}P/\mathrm{d}A)\mathrm{d}t$	J m^{-2}	7
Einstein coefficient,				8, 9
spontaneous emission	A_{ij}	$\mathrm{d}N_j/\mathrm{d}t = -\sum_i A_{ij} N_j$	s^{-1}	
stimulated or induced emission	B_{ij}	$\mathrm{d}N_j/\mathrm{d}t = -\sum_i \rho_{\tilde{\nu}}(\tilde{\nu}_{ij}) B_{ij} N_j$	s kg^{-1}	
absorption	B_{ji}	$\mathrm{d}N_i/\mathrm{d}t = -\sum_j \rho_{\tilde{\nu}}(\tilde{\nu}_{ij}) B_{ji} N_i$	s kg^{-1}	
emissivity, emittance	ε	$\varepsilon = M/M_{\text{bb}}$	1	10
Stefan-Boltzmann constant	σ	$M_{\text{bb}} = \sigma T^4$	W m^{-2} K^{-4}	10
étendue (throughput, light gathering power)	$E, (e)$	$E = A\Omega = P/L$	m^2 sr	11

(4) (continued) radiation source; Θ is the angle between the normal to the area element and the direction of observation as seen from the source.

(5) The intensity or irradiance is the radiation power per area that is received at a surface. Intensity, symbol I, is usually used in discussions involving collimated beams of light, as in applications of the Beer-Lambert law for spectrometric analysis. Intensity of electromagnetic radiation can also be defined as the modulus of the Poynting vector (see Section 2.3, p. 17 and Section 7.4, p. 148). In photochemistry the term intensity is sometimes used as an alias for radiant intensity and must not be understood as an irradiance, for which the symbol E is preferred [60].

(6) Spectral quantities may be defined with respect to frequency ν, wavelength λ, or wavenumber $\tilde{\nu}$; see the entry for spectral radiant energy density in this table.

(7) Fluence is used in photochemistry to specify the energy per area delivered in a given time interval (for instance by a laser pulse); fluence is the time integral of the fluence rate. Sometimes distinction must be made between irradiance and fluence rate [60]; fluence rate reduces to irradiance for a light beam incident from a single direction perpendicularly to the surface. The time integral of irradiance is called radiant exposure.

(8) The indices i and j refer to individual states; $E_j > E_i$, $E_j - E_i = hc\tilde{\nu}_{ij}$, and $B_{ji} = B_{ij}$ in the defining equations. The coefficients B are defined here using energy density $\rho_{\tilde{\nu}}$ in terms of wavenumber; they may alternatively be defined using energy density in terms of frequency ρ_{ν}, in which case B has SI units m kg^{-1}, and $B_{\nu} = c_0 B_{\tilde{\nu}}$ where B_{ν} is defined using frequency and $B_{\tilde{\nu}}$ using wavenumber. The defining equations refer to the *partial* contributions to the rate of change.

(9) The relation between the Einstein coefficients A and $B_{\tilde{\nu}}$ is $A = 8\pi hc_0 \tilde{\nu}^3 B_{\tilde{\nu}}$. The Einstein stimulated absorption or emission coefficient B may also be related to the transition moment between the states i and j; for an electric dipole transition the relation is

$$B_{\tilde{\nu},ij} = \frac{8\pi^3}{3h^2 c_0 (4\pi\varepsilon_0)} \sum_\rho |\langle i|\mu_\rho|j\rangle|^2$$

in which the sum over ρ runs over the three space-fixed cartesian axes, and μ_ρ is a space-fixed component of the dipole moment operator. Again, these equations are based on a wavenumber definition of the Einstein coefficient B (i.e. $B_{\tilde{\nu}}$ rather than B_{ν}).

(10) The emittance of a sample is the ratio of the radiant excitance emitted by the sample to the radiant excitance emitted by a black body at the same temperature; M_{bb} is the latter quantity. See Chapter 5, p. 112 for the value of the Stefan-Boltzmann constant.

Name	Symbol	Definition	SI unit	Notes
resolution	$\delta\tilde{\nu}$		m^{-1}	2, 12, 13
resolving power	R	$R = \tilde{\nu}/\delta\tilde{\nu}$	1	13
free spectral range	$\Delta\tilde{\nu}$	$\Delta\tilde{\nu} = 1/2l$	m^{-1}	2, 14
finesse	f	$f = \Delta\tilde{\nu}/\delta\tilde{\nu}$	1	14
quality factor	Q	$Q = 2\pi\nu\dfrac{W}{-dW/dt}$	1	14, 15
first radiation constant	c_1	$c_1 = 2\pi h c_0{}^2$	$W\ m^2$	
second radiation constant	c_2	$c_2 = hc_0/k_B$	$K\ m$	16
transmittance, transmission factor	τ, T	$\tau = P_{tr}/P_0$	1	17, 18
absorptance, absorption factor	α	$\alpha = P_{abs}/P_0$	1	17, 18
reflectance, reflection factor	ρ, R	$\rho = P_{refl}/P_0$	1	17, 18
(decadic) absorbance	A_{10}, A	$A_{10} = -\lg(1 - \alpha_i)$	1	17–20
napierian absorbance	A_e, B	$A_e = -\ln(1 - \alpha_i)$	1	17–20
absorption coefficient,				
(linear) decadic	a, K	$a = A_{10}/l$	m^{-1}	18, 21
(linear) napierian	α	$\alpha = A_e/l$	m^{-1}	18, 21
molar (decadic)	ε	$\varepsilon = a/c = A_{10}/cl$	$m^2\ mol^{-1}$	18, 21, 22
molar napierian	κ	$\kappa = \alpha/c = A_e/cl$	$m^2\ mol^{-1}$	18, 21, 22

(11) Étendue is a characteristic of an optical instrument. It is a measure of the light gathering power, i.e. the power transmitted per radiance of the source. A is the area of the source or image stop; Ω is the solid angle accepted from each point of the source by the aperture stop.

(12) The precise definition of resolution depends on the lineshape, but usually resolution is taken as the full line width at half maximum intensity (FWHM) on a wavenumber, $\delta\tilde{\nu}$, or frequency, $\delta\nu$, scale. Frequently the use of resolving power, of dimension 1, is preferable.

(13) This quantity characterizes the performance of a spectrometer, or the degree to which a spectral line (or a laser beam) is monochromatic. It may also be defined using frequency ν, or wavelength λ.

(14) These quantities characterize a Fabry-Perot cavity, or a laser cavity. l is the cavity spacing, and $2l$ is the round-trip path length. The free spectral range is the wavenumber interval between successive longitudinal cavity modes.

(15) W is the energy stored in the cavity, and $-dW/dt$ is the rate of decay of stored energy. Q is also related to the linewidth of a single cavity mode: $Q = \nu/\delta\nu = \tilde{\nu}/\delta\tilde{\nu}$. Thus high Q cavities give narrow linewidths.

(16) k_B is the Boltzmann constant (see Section 2.9, p. 45 and Chapter 5, p. 111).

(17) If scattering and luminescence can be neglected, $\tau + \alpha + \rho = 1$. In optical spectroscopy internal properties (denoted by subscript i) are defined to exclude surface effects and effects of the cuvette such as reflection losses, so that $\tau_i + \alpha_i = 1$, if scattering and luminescence can be neglected. This leads to the customary form of the Beer-Lambert law, $P_{tr}/P_0 = I_{tr}/I_0 = \tau_i = 1 - \alpha_i = \exp(-\kappa cl)$. Hence $A_e = -\ln(\tau_i)$, $A_{10} = -\lg(\tau_i)$.

(18) In spectroscopy all of these quantities are commonly taken to be defined in terms of the spectral intensity, $I_{\tilde{\nu}}(\tilde{\nu})$, hence they are all regarded as functions of wavenumber $\tilde{\nu}$ (or frequency ν) across the spectrum. Thus, for example, the absorption coefficient $\alpha(\tilde{\nu})$ as a function of wavenumber $\tilde{\nu}$ defines the absorption spectrum of the sample; similarly $T(\tilde{\nu})$ defines the transmittance spectrum. Spectroscopists use $I(\tilde{\nu})$ instead of $I_{\tilde{\nu}}(\tilde{\nu})$.

(19) The definitions given here relate the absorbance A_{10} or A_e to the *internal* absorptance α_i (see

Name	Symbol	Definition	SI unit	Notes
net absorption cross section	σ_{net}	$\sigma_{net} = \kappa/N_A$	m^2	23
absorption coefficient				
integrated over $\widetilde{\nu}$	$A, \bar{A}$	$A = \int \kappa(\widetilde{\nu}) \, d\widetilde{\nu}$	$m\ mol^{-1}$	23, 24
	S	$S = A/N_A$	m	23, 24
	$\bar{S}$	$\bar{S} = (1/pl) \int \ln(I_0/I) \, d\widetilde{\nu}$	$Pa^{-1}\ m^{-2}$	23–25
integrated over $\ln \widetilde{\nu}$	Γ	$\Gamma = \int \kappa(\widetilde{\nu})\widetilde{\nu}^{-1} \, d\widetilde{\nu}$	$m^2\ mol^{-1}$	23, 24
integrated net absorption				
cross section	G_{net}	$G_{net} = \int \sigma_{net}(\widetilde{\nu})\widetilde{\nu}^{-1} \, d\widetilde{\nu}$	m^2	23, 24
absorption index,				
imaginary refractive index	k	$k = \alpha/4\pi\widetilde{\nu}$	1	26
complex refractive index	$\hat{n}$	$\hat{n} = n + ik$	1	
molar refraction,				
molar refractivity	R	$R = \left(\dfrac{n^2-1}{n^2+2}\right) V_m$	$m^3\ mol^{-1}$	
angle of optical rotation	α		1, rad	27
specific optical rotatory power	$[\alpha]_\lambda^\theta$	$[\alpha]_\lambda^\theta = \alpha/\gamma l$	$rad\ m^2\ kg^{-1}$	27
molar optical rotatory power	α_m	$\alpha_m = \alpha/cl$	$rad\ m^2\ mol^{-1}$	27

(19) (continued) note 17). However the subscript i on the absorptance α is often omitted. Experimental data must include corrections for reflections, scattering and luminescence, if the absorbance is to have an absolute meaning. In practice the absorbance is measured as the logarithm of the ratio of the light transmitted through a reference cell (with solvent only) to that transmitted through a sample cell.

(20) In reference [57] the symbol A is used for decadic absorbance, and B for napierian absorbance.

(21) l is the absorbing path length, and c is the amount (of substance) concentration.

(22) The molar decadic absorption coefficient ε is sometimes called the "extinction coefficient" in the published literature. Unfortunately numerical values of the "extinction coefficient" are often quoted without specifying units; the absence of units usually means that the units are $mol^{-1}\ dm^3\ cm^{-1}$ (see also [61]). The word "extinction" should properly be reserved for the sum of the effects of absorption, scattering, and luminescence.

(23) Note that these quantities give the net absorption coefficient κ, the net absorption cross section σ_{net}, and the net values of $A, S, \bar{S}, \Gamma$, and G_{net}, in the sense that they are the sums of effects due to absorption and induced emission (see the discussion in Section 2.7.1, p. 38).

(24) The definite integral defining these quantities may be specified by the limits of integration in parentheses, e.g. $G(\widetilde{\nu}_1, \widetilde{\nu}_2)$. In general the integration is understood to be taken over an absorption line or an absorption band. $A, \bar{S}$, and Γ are measures of the strength of the band in terms of amount concentration; $G_{net} = \Gamma/N_A$ and $S = A/N_A$ are corresponding molecular quantities. For a single spectral line the relation of these quantities to the Einstein transition probabilities is discussed in Section 2.7.1, p. 38. The symbol $\bar{A}$ may be used for the integrated absorption coefficient A when there is a possibility of confusion with the Einstein spontaneous emission coefficient A_{ij}.

The integrated absorption coefficient of an electronic transition is often expressed in terms of the oscillator strength or "f-value", which is dimensionless, or in terms of the Einstein transition probability A_{ij} between the states involved, with SI unit s^{-1}. Whereas A_{ij} has a simple and universally accepted meaning (see note 9, p. 35), there are differing uses of f. A common practical conversion is given by the equation

$$f_{ij} = [(4\pi\varepsilon_0)\, m_e c_0/8\pi^2 e^2]\lambda^2 A_{ij} \qquad \text{or} \qquad f_{ij} \approx \left(1.4992 \times 10^{-14}\right) \left(A_{ij}/s^{-1}\right) (\lambda/nm)^2$$

in which λ is the transition wavelength, and i and j refer to individual states. For strongly allowed electronic transitions f is of order unity.

2.7.1 Quantities and symbols concerned with the measurement of absorption intensity

In most experiments designed to measure the intensity of spectral absorption, the measurement gives the net absorption due to the effects of absorption from the lower energy level m to the upper energy level n, minus induced emission from n to m. Since the populations depend on the temperature, so does the measured net absorption. This comment applies to all the quantities defined in the table to measure absorption intensity, although for transitions where $hc_0\tilde{\nu} \gg kT$ the temperature dependence is small and for $\tilde{\nu} > 1000$ cm^{-1} at ordinary temperatures induced emission can generally be neglected.

In a more fundamental approach one defines the cross section $\sigma_{ji}(\tilde{\nu})$ for an induced radiative transition from the state i to the state j (in either absorption or emission). For an ideal absorption experiment with only the lower state i populated the integrated absorption cross section for the transition $j \leftarrow i$ is given by

$$G_{ji} = \int \sigma_{ji}(\tilde{\nu})\tilde{\nu}^{-1}\mathrm{d}\tilde{\nu} = \int \sigma_{ji}(\nu)\nu^{-1}\mathrm{d}\nu$$

If the upper and lower energy levels are degenerate, the observed line strength is given by summing over transitions between all states i in the lower energy level m and all states j in the upper energy level n, multiplying each term by the fractional population p_i in the appropriate initial state. Neglecting induced emission this gives

$$G_{\mathrm{net}}(n \leftarrow m) = \sum_{i,j} p_i G_{ji}$$

(Notes continued)

(25) The quantity $\bar{S}$ is only used for gases; it is defined in a manner similar to A, except that the partial pressure p of the gas replaces the concentration c. At low pressures $p_i \approx c_i RT$, so that $\bar{S}$ and A are related by the equation $\bar{S} \approx A/RT$. Thus if $\bar{S}$ is used to report line or band intensities, the temperature should be specified. I_0 is the incident, I the transmitted intensity, thus $\ln(I_0/I) = -\ln(I/I_0) = -\ln(1 - P_{\mathrm{abs}}/P_0) = A_{\mathrm{e}}$ (see also notes 17 and 19, p. 36).

(26) α in the definition is the napierian absorption coefficient.

(27) The sign convention for the angle of optical rotation is as follows: α is positive if the plane of polarization is rotated clockwise as viewed looking towards the light source. If the rotation is anti clockwise, then α is negative. The optical rotation due to a solute in solution may be specified by a statement of the type

$$\alpha(589.3 \text{ nm}, 20\,^\circ\text{C}, \text{sucrose}, 10 \text{ g dm}^{-3} \text{ in H}_2\text{O}, 10 \text{ cm path}) = +0.6647^\circ$$

The same information may be conveyed by quoting either the specific optical rotatory power $\alpha/\gamma l$, or the molar optical rotatory power α/cl, where γ is the mass concentration, c is the amount (of substance) concentration, and l is the path length. Most tabulations give the specific optical rotatory power, denoted $[\alpha]_\lambda^\theta$. The wavelength of light used λ (frequently the sodium D line) and the Celsius temperature θ are conventionally written as a subscript and superscript to the specific rotatory power $[\alpha]$. For pure liquids and solids $[\alpha]_\lambda^\theta$ is similarly defined as $[\alpha]_\lambda^\theta = \alpha/\rho l$, where ρ is the mass density.

Specific optical rotatory powers are customarily called *specific rotations*, and are unfortunately usually quoted without units. The absence of units may usually be taken to mean that the units are $^\circ$ cm^3 g^{-1} dm^{-1} for pure liquids and solutions, or $^\circ$ cm^3 g^{-1} mm^{-1} for solids, where $^\circ$ is used as a symbol for degrees of plane angle.

If induced emission is significant then the net integrated cross section becomes

$$G_{\text{net}}\left(n \leftarrow m\right) = \sum_{i,j} \left(p_i - p_j\right) G_{ji} = \left(\frac{p_m}{d_m} - \frac{p_n}{d_n}\right) \sum_{i,j} G_{ji}$$

Here p_i and p_j denote the fractional populations of states i and j ($p_i = \exp\{-E_i/kT\}/q$ in thermal equilibrium, where q is the partition function); p_m and p_n denote the corresponding fractional populations of the energy levels, and d_m and d_n the degeneracies ($p_i = p_m/d_m$, etc.). The absorption intensity G_{ji}, and the Einstein coefficients A_{ij} and B_{ji}, are fundamental measures of the line strength between the individual states i and j; they are related to each other by the general equations

$$G_{ji} = hB_{\widetilde{\nu},ji} = (h/c_0)\, B_{\nu,ji} = A_{ij}/(8\pi c_0 \widetilde{\nu}^3)$$

Finally, for an electric dipole transition these quantities are related to the square of the transition moment by the equation

$$G_{ji} = hB_{\widetilde{\nu},ji} = A_{ij}/(8\pi c_0 \widetilde{\nu}^3) = \frac{8\pi^3}{3hc_0(4\pi\varepsilon_0)}\, |M_{ji}|^2$$

where the transition moment M_{ji} is given by

$$|M_{ji}|^2 = \sum_\rho |\langle i\, |\mu_\rho|\, j\rangle|^2$$

Here the sum is over the three space-fixed cartesian axes and μ_ρ is a space-fixed component of the electric dipole moment. Inserting numerical values for the constants the relation between G_{ji} and M_{ji} may be expressed in a practical representation as

$$\left(G_{ji}/\text{pm}^2\right) \approx 41.6238\, |M_{ji}/\text{D}|^2$$

where $1\,\text{D} \approx 3.335\,641 \times 10^{-30}$ C m (D is the symbol of the debye) [62].

Net integrated absorption band intensities are usually characterized by one of the quantities $A, S, \bar{S}, \Gamma$, or G_{net} as defined in the table. The relation between these quantities is given by the (approximate) equations

$$G_{\text{net}} = \Gamma/N_A \approx A/(\widetilde{\nu}_0 N_A) \approx S/\widetilde{\nu}_0 \approx \bar{S}\left(kT/\widetilde{\nu}_0\right)$$

However, only the first equality is exact. The relation to A, $\bar{S}$ and S involves dividing by the band centre wavenumber $\widetilde{\nu}_0$ for a band, to correct for the fact that A, $\bar{S}$ and S are obtained by integrating over wavenumber rather than integrating over the logarithm of wavenumber used for G_{net} and Γ. This correction is only approximate for a band (although negligible error is involved for single-line intensities in gases). The relation to $\bar{S}$ involves the assumption that the gas is ideal (which is approximately true at low pressures), and also involves the temperature. Thus the quantities Γ and G_{net} are most simply related to more fundamental quantities such as the Einstein transition probabilities and the transition moment, and are the preferred quantities for reporting integrated line or band intensities.

The situation is further complicated by some authors using the symbol S for any of the above quantities, particularly for any of the quantities here denoted A, S and $\bar{S}$. It is therefore particularly important to define quantities and symbols used in reporting integrated intensities.

For transitions between individual states any of the more fundamental quantities G_{ji}, $B_{\widetilde{\nu},ji}$, A_{ij}, or $|M_{ji}|$ may be used; the relations are as given above, and are exact. The integrated absorption coefficient A should not be confused with the Einstein coefficient A_{ij} (nor with absorbance, for which the symbol A is also used). Where such confusion might arise, we recommend writing $\bar{A}$ for the band intensity expressed as an integrated absorption coefficient over wavenumber.

The SI unit and commonly used units of A, S, $\bar{S}$, Γ and G are as in the table below. Also given in the table are numerical transformation coefficients in commonly used units, from A, S, $\bar{S}$, and Γ to G_{net}.

Quantity	SI unit	Common unit	Transformation coefficient
$A, \bar{A}$	m mol^{-1}	km mol^{-1}	$(G/\mathrm{pm}^2) = 16.605\ 40\ \dfrac{A/(\mathrm{km\ mol}^{-1})}{\widetilde{\nu}_0/\ \mathrm{cm}^{-1}}$
$\bar{S}$	Pa^{-1} m^{-2}	atm^{-1} cm^{-2}	$(G/\mathrm{pm}^2) = 1.362\ 603 \times 10^{-2}\ \dfrac{(\bar{S}/\mathrm{atm}^{-1}\ \mathrm{cm}^{-2})\,(T/\mathrm{K})}{(\widetilde{\nu}_0/\mathrm{cm}^{-1})}$
S	m	cm	$(G/\mathrm{pm}^2) = 10^{20}\ \dfrac{(S/\mathrm{cm})}{\widetilde{\nu}_0/\mathrm{cm}^{-1}}$
Γ	m^2 mol^{-1}	cm^2 mol^{-1}	$(G/\mathrm{pm}^2) = 1.660\ 540 \times 10^{-4}\ (\Gamma/\mathrm{cm}^2\ \mathrm{mol}^{-1})$
G	m^2	pm^2	

Quantities concerned with spectral absorption intensity and relations among these quantities are discussed in references [62–65], and a list of published measurements of line intensities and band intensities for gas phase infrared spectra may be found in references [63–65]. For relations between spectroscopic absorption intensities and kinetic radiative transition rates see also [66].

2.7.2 Conventions for absorption intensities in condensed phases

Provided transmission measurements are accurately corrected for reflection and other losses, the absorbance, absorption coefficient, and integrated absorption coefficient, A, that are described above may be used for condensed phases. The corrections typically used are adequate for weak and medium bands in neat liquids and solids and for bands of solutes in dilute solution. In order to make the corrections accurately for strong bands in neat liquids and solids, it is necessary to determine the real and imaginary refractive indices n and k, of the sample throughout the measured spectral range. Then, the resulting n and k themselves provide a complete description of the absorption intensity in the condensed phase. For liquids, this procedure requires knowledge of n and k of the cell windows. Reflection spectra are also processed to yield n and k. For non-isotropic solids all intensity properties must be defined with respect to specific crystal axes. If the n and k spectra are known, spectra of any optical property or measured quantity of the sample may be obtained from them. Physicists prefer to use the complex relative permittivity (see Section 2.3, p. 16), $\widehat{\varepsilon}_r = \varepsilon_r{}' + \mathrm{i}\,\varepsilon_r{}'' = \mathrm{Re}\,\widehat{\varepsilon}_r + \mathrm{i}\,\mathrm{Im}\,\widehat{\varepsilon}_r$ instead of the complex refractive index $\widehat{n}$. $\varepsilon_r{}' = \mathrm{Re}\,\widehat{\varepsilon}_r$ denotes the real part and $\varepsilon_r{}'' = \mathrm{Im}\,\widehat{\varepsilon}_r$ the imaginary part of $\widehat{\varepsilon}_r$, respectively. They are related through $\widehat{\varepsilon}_r = \widehat{n}^2$, so that $\varepsilon_r' = n^2 - k^2$ and $\varepsilon_r{}'' = 2nk$.

The refractive indices and relative permittivities are properties of the bulk phase. In order to obtain information about the molecules in the liquid free from dielectric effects of the bulk, the local field that acts on the molecules, $\boldsymbol{E}_{\mathrm{loc}}$, must be determined as a function of the applied field $\boldsymbol{E}$. A simple relation is the Lorentz local field, $\boldsymbol{E}_{\mathrm{loc}} = \boldsymbol{E} + \boldsymbol{P}/3\varepsilon_0$, where $\boldsymbol{P}$ is the dielectric polarization. This local field is based on the assumption that long-range interactions are isotropic, so it is realistic only for liquids and isotropic solids.

Use of this local field gives the Lorentz-Lorenz formula (this relation is usually called the Clausius-Mossotti formula when applied to static fields) generalized to treat absorbing materials at any wavenumber

$$\frac{\widehat{\varepsilon}_r(\widetilde{\nu}) - 1}{\widehat{\varepsilon}_r(\widetilde{\nu}) + 2} = \frac{1}{3\varepsilon_0 V_m}\widehat{\alpha}_m(\widetilde{\nu})$$

Here V_m is the molar volume, and $\widehat{\alpha}_m$ is the complex molar polarizability (see Section 2.3, note 7, p. 16). The imaginary part α_m'' of the complex molar polarizability describes the absorption by molecules in the liquid, corrected for the long-range isotropic dielectric effects but influenced by the anisotropic environment formed by the first few nearest neighbor molecules. The real part α_m' is the molar polarizability (see sections 2.3 and 2.5) in the limit of infinite frequency.

The integrated absorption coefficient of a molecular absorption band in the condensed phase is described by [67,68] (see note 1, below)

$$C_j = \frac{1}{4\pi\varepsilon_0}\int_{\text{band } j} \widetilde{\nu}\alpha_m''(\widetilde{\nu})\mathrm{d}\widetilde{\nu}$$

Theoretical analysis usually assumes that the measured band includes the transition $j \leftarrow i$ with band centre wavenumber $\widetilde{\nu}_0$ and all of its hot-band transitions. Then

$$C_j = \frac{N_A\pi}{3hc_0(4\pi\varepsilon_0)}\widetilde{\nu}_0 g(|M_{ji}|^2)$$

where $\widetilde{\nu}_0 g(|M_{ji}|^2)$ is the population-weighted sum over all contributing transitions of the wavenumber times the square of the electric dipole moment of the transition (g is in general a temperature dependent function which includes effects from the inhomogeneous band structure). The traditional relation between the gas and liquid phase values of the absorption coefficient of a given band j is the Polo-Wilson equation [69]

$$A_{\text{liq}} = \frac{(\bar{n}^2 + 2)^2}{9\bar{n}}A_{\text{gas}}$$

where $\bar{n}$ is the estimated average value of n through the absorption band. This relation is valid under the following conditions:

(i) bands are sufficiently weak, i.e. $2nk + k^2 \ll n^2$;

(ii) the double harmonic approximation yields a satisfactory theoretical approximation of the band structure; in this case the function g is nearly temperature independent and the gas and liquid phase values of g are equal;

(iii) $\widetilde{\nu}_0|M_{ji}|^2$ are identical in the gas and liquid phases.

A more recent and more general relation that requires only the second and the third of these conditions is $A_{\text{gas}} = 8\pi^2 C_j$. It follows that for bands that meet all three of the conditions

$$A_{\text{liq}} = 8\pi^2 C_j \frac{(\bar{n}^2 + 2)^2}{9\bar{n}}$$

This relation shows that, while the refractive index n is automatically incorporated in the integrated absorption coefficient C_j for a liquid, the traditional absorption coefficient A needs to be further corrected for permittivity or refraction.

---

(1) The SI unit of this quantity is m mol^{-1}. In the Gaussian system the definition of C_j as $C_j = \int_{\text{band } j} \widetilde{\nu}\alpha_m''(\widetilde{\nu})\mathrm{d}\widetilde{\nu}$ is often adopted, with unit cm mol^{-1}.

2.8 SOLID STATE

The quantities and their symbols given here have been selected from more extensive lists of IUPAP [4] and ISO [5.n]. See also the *International Tables for Crystallography*, Volume A [70].

Name	Symbol	Definition	SI unit	Notes
lattice vector,				
Bravais lattice vector	$\boldsymbol{R}, \boldsymbol{R}_0$		m	
fundamental translation	$\boldsymbol{a}_1; \boldsymbol{a}_2; \boldsymbol{a}_3,$	$\boldsymbol{R} = n_1\boldsymbol{a}_1 + n_2\boldsymbol{a}_2 + n_3\boldsymbol{a}_3$	m	1
vectors for the crystal	$\boldsymbol{a}; \boldsymbol{b}; \boldsymbol{c}$			
lattice				
(angular) fundamental				
translation vectors	$\boldsymbol{b}_1; \boldsymbol{b}_2; \boldsymbol{b}_3,$	$\boldsymbol{a}_i \cdot \boldsymbol{b}_k = 2\pi\delta_{ik}$	m^{-1}	2
for the reciprocal lattice	$\boldsymbol{a}^*; \boldsymbol{b}^*; \boldsymbol{c}^*$			
(angular) reciprocal	$\boldsymbol{G}$	$\boldsymbol{G} = h_1\boldsymbol{b}_1 + h_2\boldsymbol{b}_2 + h_3\boldsymbol{b}_3 =$	m^{-1}	1, 3
lattice vector		$h_1\boldsymbol{a}^* + h_2\boldsymbol{b}^* + h_3\boldsymbol{c}^*$		
		$\boldsymbol{G} \cdot \boldsymbol{R} = 2\pi m = 2\pi \sum_i n_i h_i$		
unit cell lengths	$a; b; c$		m	
unit cell angles	$\alpha; \beta; \gamma$		rad, 1	
reciprocal unit cell lengths	$a^*; b^*; c^*$		m^{-1}	
reciprocal unit cell angles	$\alpha^*; \beta^*; \gamma^*$		rad, 1	
fractional coordinates	$x; y; z$	$x = X/a$	1	4
atomic scattering factor	f	$f = E_a/E_e$	1	5
structure factor	$F(h, k, l)$	$F = \sum_{n=1}^{N} f_n e^{2\pi i(hx_n+ky_n+lz_n)}$	1	6
with indices h, k, l				
lattice plane spacing	d		m	
Bragg angle	θ	$n\lambda = 2d\sin\theta$	rad, 1	7
order of reflection	n		1	
order parameters,				
short range	σ		1	
long range	s		1	
Burgers vector	$\boldsymbol{b}$		m	
particle position vector	$\boldsymbol{r}_j, \boldsymbol{R}_j$		m	8
equilibrium position	$\boldsymbol{R}_0$		m	
vector of an ion				
displacement vector	$\boldsymbol{u}$	$\boldsymbol{u} = \boldsymbol{R} - \boldsymbol{R}_0$	m	
of an ion				
Debye-Waller factor	B, D	$D = e^{-2<(\boldsymbol{q}\cdot\boldsymbol{u})^2>}$	1	9

(1) n_1, n_2 and n_3 are integers. a, b and c are also called the lattice constants.

(2) Reciprocal lattice vectors are sometimes defined by $\boldsymbol{a}_i \cdot \boldsymbol{b}_k = \delta_{ik}$.

(3) m is an integer with $m = n_1 h + n_2 k + n_3 l$.

(4) X denotes the coordinate of dimension length.

(5) E_a and E_e denote the scattering amplitudes for the atom and the isolated electron, respectively.

(6) N is the number of atoms in the unit cell.

(7) λ is the wavelength of the incident radiation.

(8) To distinguish between electron and ion position vectors, lower case and capital letters are used respectively. The subscript j relates to particle j.

(9) $\hbar\boldsymbol{q}$ is the momentum transfer in the scattering of neutrons, $<>$ denotes thermal averaging.

Name	Symbol	Definition	SI unit	Notes
Debye angular wavenumber	q_D, k_D	$k_D = (C_i\, 6\pi^2)^{1/3}$	m^{-1}	10
Debye angular frequency	ω_D	$\omega_D = k_D\, c_0$	s^{-1}	10
Debye frequency	ν_D	$\nu_D = \omega_D/2\pi$	s^{-1}	
Debye wavenumber	$\tilde{\nu}_D$	$\tilde{\nu}_D = \nu_D/c_0$	m^{-1}	10
Debye temperature	Θ_D	$\Theta_D = h\nu_D/k_B$	K	
Grüneisen parameter	γ, Γ	$\gamma = \alpha V/\kappa C_V$	1	11
Madelung constant	$\alpha, \mathcal{M}$	$E_{coul} = \dfrac{\alpha N_A z_+ z_- e^2}{4\pi\varepsilon_0 R_0}$	1	12
density of states	N_E	$N_E = \mathrm{d}N(E)/\mathrm{d}E$	$J^{-1}\ m^{-3}$	13
(spectral) density of vibrational modes	N_ω, g	$N_\omega = \mathrm{d}N(\omega)/\mathrm{d}\omega$	$s\ m^{-3}$	14
resistivity tensor	$\boldsymbol{\rho}, (\rho_{ik})$	$\boldsymbol{E} = \boldsymbol{\rho} \cdot \boldsymbol{j}$	$\Omega\ m$	15
conductivity tensor	$\boldsymbol{\sigma}, (\sigma_{ik})$	$\boldsymbol{j} = \boldsymbol{\sigma} \cdot \boldsymbol{E}$	$S\ m^{-1}$	15
residual resistivity	ρ_R		$\Omega\ m$	
thermal conductivity tensor	λ_{ik}	$\boldsymbol{J}_q = -\boldsymbol{\lambda} \cdot \boldsymbol{\nabla} T$	$W\ m^{-1}\ K^{-1}$	15
relaxation time	τ	$\tau = l/v_F$	s	16
Lorenz coefficient	L	$L = \lambda/\sigma T$	$V^2\ K^{-2}$	17
Hall coefficient	A_H, R_H	$\boldsymbol{E} = \rho\boldsymbol{j} + R_H(\boldsymbol{B} \times \boldsymbol{j})$	$m^3\ C^{-1}$	
thermoelectric force	E		V	18
Peltier coefficient	Π	$E = \Pi\dfrac{\Delta T}{T}$	V	18
Thomson coefficient	$\mu, (\tau)$	$\mu = \Pi/T$	$V\ K^{-1}$	
number density, number concentration	C, n, p		m^{-3}	19
band gap energy	E_g		J	20
donor ionization energy	E_d		J	20
acceptor ionization energy	E_a		J	20
Fermi energy	E_F, ε_F	$\varepsilon_F = \lim\limits_{T \to 0} \mu$	J	20
work function, electron work function	Φ	$\Phi = E_\infty - E_F$	J	21
angular wave vector, propagation vector	$\boldsymbol{k}, \boldsymbol{q}$	$k = 2\pi/\lambda$	m^{-1}	22
Bloch function	$u_k(\boldsymbol{r})$	$\psi(\boldsymbol{r}) = u_k(\boldsymbol{r})\exp(\mathrm{i}\boldsymbol{k}\cdot\boldsymbol{r})$	$m^{-3/2}$	23
charge density of electrons	ρ	$\rho(\boldsymbol{r}) = -e\psi^*(\boldsymbol{r})\psi(\boldsymbol{r})$	$C\ m^{-3}$	23, 24
effective mass	m^*		kg	25
mobility	μ	$\boldsymbol{v}_{drift} = \mu\boldsymbol{E}$	$m^2\ V^{-1}\ s^{-1}$	25
mobility ratio	b	$b = \mu_n/\mu_p$	1	
diffusion coefficient	D	$\boldsymbol{j} = -D\boldsymbol{\nabla} C$	$m^2\ s^{-1}$	25, 26
diffusion length	L	$L = \sqrt{D\tau}$	m	25, 26
characteristic (Weiss) temperature	θ, θ_W		K	
Curie temperature	T_C		K	
Néel temperature	T_N		K	

(10) C_i is the ion density, c_0 is the speed of light in vacuum. q_d is equal to 2π times the inverse of the Debye cut-off wavelength of the elastic lattice wave.

(11) α is the cubic expansion coefficient, V the volume, κ the isothermal compressibility, and C_V the heat capacity at constant volume.

(12) E_{coul} is the electrostatic interaction energy per mole of ion pairs with charges $z_+ e$ and $-z_- e$.

2.8.1 Symbols for planes and directions in crystals

Miller indices of a crystal face, or of a single net plane	(hkl) or $(h_1 h_2 h_3)$
indices of the Bragg reflection from the set of parallel net planes (hkl)	hkl or $h_1 h_2 h_3$
indices of a set of all symmetrically equivalent crystal faces, or net planes	$\{hkl\}$ or $\{h_1 h_2 h_3\}$
indices of a lattice direction (zone axis)	$[uvw]$
indices of a set of symmetrically equivalent lattice directions	$< uvw >$

In each of these cases, when the letter symbol is replaced by numbers it is customary to omit the commas. For a single plane or crystal face, or a specific direction, a negative number is indicated by a bar over the number.

Example $(\bar{1}10)$ denotes the parallel planes $h = -1$, $k = 1$, $l = 0$.

(i) Crystal lattice symbols

primitive	P
face-centred	F
body-centred	I
base-centred	A;B;C
rhombohedral	R

(ii) Hermann-Mauguin symbols of symmetry operations

Operation	Symbol	Examples
n-fold rotation	n	1; 2; 3; 4; 6
n-fold inversion	$\bar{n}$	$\bar{1}$; $\bar{2}$; $\bar{3}$; $\bar{4}$; $\bar{6}$
n-fold screw	n_k	2_1; 3_1; 3_2;...
reflection	m	
glide	$a; b; c; n; d$	

(Notes continued)

(13) $N(E)$ is the total number of states of electronic energy less than E, divided by the volume.

(14) $N(\omega)$ is the total number of vibrational modes with circular frequency less than ω, divided by the volume.

(15) Tensors may be replaced by their corresponding scalar quantities in isotropic media. $\boldsymbol{J}_q$ is the heat flux vector or thermal current density.

(16) The definition applies to electrons in metals; l is the mean free path, and v_F is the electron velocity on the Fermi sphere.

(17) λ and σ are the thermal and electrical conductivities in isotropic media.

(18) The substances to which the symbol applies are denoted by subscripts. The thermoelectric force is an electric potential difference induced by the gradient of the chemical potential.

(19) Specific number densities are denoted by subscripts: for electrons n_n, n_-, (n); for holes n_p, n_+, (p); for donors n_d; for acceptors n_a; for the intrinsic number density n_i ($n_i{}^2 = n_+ n_-$).

(20) The commonly used unit for this quantity is eV. μ is the chemical potential per entity.

(21) E_∞ is the electron energy at rest at infinite distance [71].

(22) $\boldsymbol{k}$ is used for particles, $\boldsymbol{q}$ for phonons. Here, λ is the wavelength.

(23) $\psi(\boldsymbol{r})$ is a one-electron wavefunction.

(24) The total charge density is obtained by summing over all electrons.

(25) Subscripts n and p or $-$ and $+$ may be added to denote electrons and holes, respectively.

(26) $\boldsymbol{j}$ is the particle flux density. D is the diffusion coefficient and τ the lifetime.

2.9 STATISTICAL THERMODYNAMICS

The names and symbols given here are in agreement with those recommended by IUPAP [4] and by ISO [5.h].

Name	Symbol	Definition	SI unit	Notes
number of entities	N		1	
number density of entities, number concentration	C	$C = N/V$	m^{-3}	
Avogadro constant	L, N_A	$L = N/n$	mol^{-1}	1
Boltzmann constant	k, k_B		$J\ K^{-1}$	
(molar) gas constant	R	$R = Lk$	$J\ K^{-1}\ mol^{-1}$	
molecular position vector	$\boldsymbol{r}(x, y, z)$		m	
molecular velocity vector	$\boldsymbol{c}(c_x, c_y, c_z),$ $\boldsymbol{u}(u_x, u_y, u_z),$ $\boldsymbol{v}(v_x, v_y, v_z)$	$\boldsymbol{c} = \mathrm{d}\boldsymbol{r}/\mathrm{d}t$	$m\ s^{-1}$	
molecular momentum vector	$\boldsymbol{p}(p_x, p_y, p_z)$	$\boldsymbol{p} = m\boldsymbol{c}$	$kg\ m\ s^{-1}$	2
velocity distribution function	$f(c_x)$	$f = \left(\dfrac{m}{2\pi kT}\right)^{1/2} \exp\left(-\dfrac{mc_x^{\ 2}}{2kT}\right)$	$m^{-1}\ s$	2
speed distribution function	$F(c)$	$F = 4\pi c^2\left(\dfrac{m}{2\pi kT}\right)^{3/2} \exp\left(-\dfrac{mc^2}{2kT}\right)$	$m^{-1}s$	2
average speed	$\bar{c}, \bar{u}, \bar{v},$ $\langle c\rangle, \langle u\rangle, \langle v\rangle$	$\bar{c} = \int cF(c)\ \mathrm{d}c$	ms^{-1}	
generalized coordinate	q		(varies)	3
generalized momentum	p	$p = \partial L/\partial \dot{q}$	(varies)	3
volume in phase space	Ω	$\Omega = (1/h)\int p\ \mathrm{d}q$	1	
probability	P, p		1	
statistical weight, degeneracy	g, d, W, ω, β		1	4
(cumulative) number of states	W, N	$W(E) = \sum_i H(E - E_i)$	1	5, 6
density of states	$\rho(E)$	$\rho(E) = \mathrm{d}W(E)/\mathrm{d}E$	J^{-1}	
partition function, sum over states, single molecule	q, z	$q = \sum_i g_i \exp(-E_i/kT)$	1	6
canonical ensemble, (system, or assembly)	Q, Z	$Q = \sum_i g_i \exp(-E_i/kT)$	1	6

(1) n is the amount of substance (or the chemical amount, enplethy). While the symbol N_A is used to honour Amedeo Avogadro, the symbol L is used in honour of Josef Loschmidt.
(2) m is the mass of the particle.
(3) If q is a length then p is a momentum. In the definition of p, L denotes the Lagrangian.
(4) β is sometimes used for a spin statistical weight and the degeneracy is also called polytropy. It is the number of linearly independent energy eigenfunctions for the same energy.
(5) $H(x)$ is the Heaviside function (see Section 4.2, p. 107), W or $W(E)$ is the total number of quantum states with energy less than E.
(6) E_i denotes the energy of the ith level of a molecule or quantum system under consideration and g_i denotes its degeneracy.

Name	Symbol	Definition	SI units	Notes
microcanonical ensemble partition function, sum over states,	Ω, z, Z	$Z = \sum_{i=1}^{Z} 1$	1	
grand canonical ensemble	Ξ		1	
symmetry number	σ, s		1	
reciprocal energy parameter to replace temperature	β	$\beta = 1/kT$	J^{-1}	
characteristic temperature	Θ, θ		K	7
absolute activity	λ	$\lambda_B = \exp(\mu_B/RT)$	1	8
density operator	$\widehat{\rho}, \widehat{\sigma}$	$\widehat{\rho} = \sum_k p_k |\Psi_k\rangle\langle\Psi_k|$	1	9
density matrix	$\boldsymbol{P}, \boldsymbol{\rho}$	$\boldsymbol{P} = \{P_{mn}\}$	1	10
element	P_{mn}, ρ_{mn}	$P_{mn} = \langle\phi_m|\widehat{\rho}|\phi_n\rangle$	1	10
statistical entropy	S	$S = -k \sum_i p_i \ln p_i$	$J\ K^{-1}$	11

(7) Particular characteristic temperatures are denoted with subscripts, e.g. rotational $\Theta_r = hc\widetilde{B}/k$, vibrational $\Theta_v = hc\widetilde{\nu}/k$, Debye $\Theta_D = hc\widetilde{\nu}_D/k$, Einstein $\Theta_E = hc\widetilde{\nu}_E/k$. Θ is to be preferred over θ to avoid confusion with Celsius temperature.

(8) The definition applies to entities B. μ_B is the chemical potential (see Section 2.11, p. 57).

(9) $|\Psi_k\rangle$ refers to the quantum state k of the system and p_k to the probability of this state in an ensemble. If $p_k = 1$ for a given state k one speaks of a pure state, otherwise of a mixture.

(10) The density matrix $\boldsymbol{P}$ is defined by its matrix elements P_{mn} in a set of basis states ϕ_m. Alternatively, one can write $P_{mn} = \sum_k p_k\, c_m^{(k)}\, c_n^{(k)\,*}$, where $c_m^{(k)}$ is the (complex) coefficient of ϕ_m in the expansion of $|\Psi_k\rangle$ in the basis states $\{\phi_i\}$.

(11) In the expression for the statistical entropy, k is the Boltzmann constant and p_i the average population or probability for a quantum level. The equilibrium (maximum) entropy for a microcanonical ensemble results then as $S = k \ln Z$. A variety of other ensembles beyond the microcanonical, canonical, or grand canonical can be defined.

2.10 GENERAL CHEMISTRY

The symbols given by IUPAP [4] and by ISO [5.d,5.h] are in agreement with the recommendations given here.

Name	Symbol	Definition	SI unit	Notes
number of entities (e.g. molecules, atoms, ions, formula units)	N		1	
amount of substance, amount (chemical amount)	n	$n_B = N_B/L$	mol	1, 2
Avogadro constant	L, N_A		mol^{-1}	3
mass of atom, atomic mass	m_a, m		kg	
mass of entity (molecule, formula unit)	m, m_f		kg	4
atomic mass constant	m_u	$m_u = m_a(^{12}C)/12$	kg	5
molar mass	M	$M_B = m/n_B$	$kg\ mol^{-1}$	2, 6, 7
molar mass constant	M_u	$M_u = m_u N_A$	$g\ mol^{-1}$	7, 8
relative molecular mass, (relative molar mass, molecular weight)	M_r	$M_r = m_f/m_u$	1	8
relative atomic mass, (atomic weight)	A_r	$A_r = m_a/m_u$	1	8
molar volume	V_m	$V_{m,B} = V/n_B$	$m^3\ mol^{-1}$	2, 6, 7
mass fraction	w	$w_B = m_B/\sum_i m_i$	1	2

(1) The words "of substance" may be replaced by the specification of the entity, e.g. "amount of oxygen atoms" or "amount of oxygen (or dioxygen, O_2) molecules". Note that "amount of oxygen" is ambiguous and should be used only if the meaning is clear from the context (see also the discussion in Section 2.10.1 (v), p. 53).

Example When the amount of O_2 is equal to 3 mol, $n(O_2) = 3$ mol, the amount of $(1/2)\ O_2$ is equal to 6 mol, and $n((1/2)\ O_2) = 6$ mol. Thus $n((1/2)\ O_2) = 2n(O_2)$.

(2) The definition applies to entities B which should always be indicated by a subscript or in parentheses, e.g. n_B or $n(B)$. When the chemical composition is written out, parentheses should be used, $n(O_2)$.

(3) The symbol N_A is used to honour Amedeo Avogadro, the symbol L is used in honour of Josef Loschmidt.

(4) Note that "formula unit" is not a unit (see examples in Section 2.10.1 (iii), p. 50).

(5) m_u is equal to the unified atomic mass unit, with symbol u, i.e. $m_u = 1$ u (see Section 3.7, p. 92). The dalton, symbol Da, is used as an alternative name for the unified atomic mass unit.

(6) The definition applies to pure substance, where m is the total mass and V is the total volume. However, corresponding quantities may also be defined for a mixture as m/n and V/n, where $n = \sum_i n_i$. These quantities are called the mean molar mass and the mean molar volume respectively.

(7) These names, which include the word "molar", unfortunately use the name of a *unit* in the description of a *quantity*, which in principle is to be avoided.

(8) For historical reasons the terms "molecular weight" and "atomic weight" are still used. For molecules M_r is the relative molecular mass or "molecular weight". For atoms M_r is the relative atomic mass or "atomic weight", and the symbol A_r may be used. M_r may also be called the relative molar mass, $M_{r,B} = M_B/M_u$, where $M_u = 1$ g mol^{-1}. The standard atomic weights are listed in Section 6.2, p. 117.

Name	Symbol	Definition	SI unit	Notes
volume fraction	ϕ	$\phi_B = V_B / \sum_i V_i$	1	2, 9
mole fraction, amount-of-substance fraction, amount fraction	x, y	$x_B = n_B / \sum_i n_i$	1	2, 10
(total) pressure	$p, (P)$		Pa	2, 11
partial pressure	p_B	$p_B = y_B\, p$	Pa	12
mass concentration, (mass density)	γ, ρ	$\gamma_B = m_B / V$	kg m^{-3}	2, 13, 14
number concentration, number density of entities	C, n	$C_B = N_B / V$	m^{-3}	2, 13, 15
amount concentration, concentration	$c, [B]$	$c_B = n_B / V$	mol m^{-3}	2, 13, 16
solubility	s	$s_B = c_B$ (saturated solution)	mol m^{-3}	2, 17
molality	m, b	$m_B = n_B / m_A$	mol kg^{-1}	2, 7, 14
surface concentration	Γ	$\Gamma_B = n_B / A$	mol m^{-2}	2, 18
stoichiometric number	ν		1	19
extent of reaction, advancement	ξ	$\xi = (n_B - n_{B,0}) / \nu_B$	mol	19
degree of reaction	α	$\alpha = \xi / \xi_{max}$	1	20

(9) Here, V_B and V_i are the volumes of appropriate components prior to mixing. As other definitions are possible, e.g. ISO 31 [5], the term should not be used in accurate work without spelling out the definition.

(10) For condensed phases x is used, and for gaseous mixtures y may be used [72].

(11) Pressures are often expressed in the non-SI unit bar, where 1 bar = 10^5 Pa. The standard pressure $p^{\ominus}$ = 1 bar = 10^5 Pa (see Section 2.11.1 (v), p. 62, Section 7.2, p. 138, and the conversion table on p. 233). Pressures are often expressed in millibar or hectopascal, where 1 mbar = 10^{-3} bar = 100 Pa = 1 hPa.

(12) The symbol and the definition apply to molecules B, which should be specified. In real (non-ideal) gases, care is required in defining the partial pressure.

(13) V is the volume of the mixture. Quantities that describe compositions of mixtures can be found in [72].

(14) In this definition the symbol m is used with two different meanings: m_B denotes the *molality* of solute B, m_A denotes the *mass* of solvent A (thus the unit mol kg^{-1}). This confusion of notation is avoided by using the symbol b for molality. A solution of molality 1 mol kg^{-1} is occasionally called a 1 molal solution, denoted 1 m solution; however, the symbol m must not be treated as a symbol for the unit mol kg^{-1} in combination with other units.

(15) The term number concentration and symbol C is preferred for mixtures. Care must be taken not to use the symbol n where it may be mistakenly interpreted to denote amount of substance.

(16) "Amount concentration" is an abbreviation of "amount-of-substance concentration". (The Clinical Chemistry Division of IUPAC recommends that amount-of-substance concentration be abbreviated to "substance concentration" [14,73].) The word "concentration" is normally used alone where there is no risk of confusion, as in conjunction with the name (or symbol) of a chemical substance, or as contrast to molality (see Section 2.13, p. 70). In polymer science the word "concentration" and the symbol c is normally used for mass concentration. In the older literature this quantity was often called *molarity*, a usage that should be avoided due to the risk of confusion with the quantity *molality*. Units commonly used for amount concentration are mol L^{-1} (or mol dm^{-3}), mmol L^{-1}, μmol L^{-1} etc., often denoted M, mM, μM etc. (pronounced molar, millimolar, micromolar).

2.10.1 Other symbols and conventions in chemistry

(i) The symbols for the chemical elements

The symbols for the chemical elements are (in most cases) derived from their Latin names and consist of one or two letters which should always be printed in Roman (upright) type. A complete list is given in Section 6.2, p. 117. The symbol is not followed by a full stop except at the end of a sentence.

Examples I, U, Pa, C

The symbols have two different meanings (which also reflects on their use in chemical formulae and equations):

(a) On a *microscopic* level they can denote an atom of the element. For example, Cl denotes a chlorine atom having 17 protons and 18 or 20 neutrons (giving a mass number of 35 or 37), the difference being ignored. Its mass is on average 35.4527 u in terrestrial samples.

(b) On a *macroscopic* level they denote a sample of the element. For example, Fe denotes a sample of iron, and He a sample of helium gas. They may also be used as a shorthand to denote the element: "Fe is one of the most common elements in the Earth's crust."

The term *nuclide* implies an atom of specified atomic number (proton number) and mass number (nucleon number). A nuclide may be specified by attaching the mass number as a left superscript to the *symbol* for the element, as in ^{14}C, or added after the *name* of the element, as in carbon-14. Nuclides having the same atomic number but different mass numbers are called isotopic nuclides or *isotopes*, as in ^{12}C, ^{14}C. If no left superscript is attached, the symbol is read as including all isotopes in natural abundance: $n(Cl) = n(^{35}Cl) + n(^{37}Cl)$. Nuclides having the same mass number but different atomic numbers are called isobaric nuclides or *isobars*: ^{14}C, ^{14}N. The atomic number may be attached as a left subscript: $^{14}_{6}C$, $^{14}_{7}N$.

The ionic charge number is denoted by a right superscript, by the sign alone when the charge number is equal to plus one or minus one.

Examples		
	Na^+	a sodium positive ion (cation)
	$^{79}Br^-$	a bromine-79 negative ion (anion, bromide ion)
	Al^{3+} or Al^{+3}	aluminium triply positive ion
	$3\ S^{2-}$ or $3\ S^{-2}$	three sulfur doubly negative ions (sulfide ions)

Al^{3+} is commonly used in chemistry and recommended by [74]. The forms Al^{+3} and S^{-2}, although widely used, are obsolete [74], as well as the old notation Al^{+++}, $S^=$, and S^{--}.

(Notes continued)

(16) (continued) Thus M is often treated as a symbol for mol L^{-1}.

(17) Solubility may also be expressed in any units corresponding to quantities that denote relative composition, such as mass fraction, amount fraction, molality, volume fraction, etc.

(18) *A* denotes the surface area.

(19) The stoichiometric number is defined through the stoichiometric equation. It is negative for reactants and positive for products. The values of the stoichiometric numbers depend on how the reaction equation is written (see Section 2.10.1 (iv), p. 52). $n_{B,0}$ denotes the value of n_B at "zero time", when $\xi = 0$ mol.

(20) ξ_{max} is the value of ξ when at least one of the reactants is exhausted. For specific reactions, terms such as "degree of dissociation" and "degree of ionization" etc. are commonly used.

The right superscript position is also used to convey other information. Excited electronic states may be denoted by an asterisk.

Examples H*, Cl*

Oxidation numbers are denoted by positive or negative Roman numerals or by zero (see also Section 2.10.1 (iv), p. 52).

Examples Mn^{VII}, manganese(VII), O^{-II}, Ni^0

The positions and meanings of indices around the symbol of the element are summarized as follows:

left superscript	mass number
left subscript	atomic number
right superscript	charge number, oxidation number, excitation
right subscript	number of atoms per entity (see Section 2.10.1 (iii) below)

(ii) Symbols for particles and nuclear reactions

proton	p, p^+	positron	e^+, β^+	triton	t
antiproton	$\bar{p}$	positive muon	μ^+	helion	h ($^3\text{He}^{2+}$)
neutron	n	negative muon	μ^-	α-particle	α ($^4\text{He}^{2+}$)
antineutron	$\bar{n}$	photon	γ	(electron) neutrino	ν_e
electron	e, e^-, β^-	deuteron	d	electron antineutrino	$\bar{\nu}_e$

Particle symbols are printed in Roman (upright) type (but see Chapter 6, p. 113). The electric charge of particles may be indicated by adding the superscript +, −, or 0; e.g. p^+, n^0, e^-, etc. If the symbols p and e are used without a charge, they refer to the positive proton and negative electron respectively. A summary of recommended names for muonium and hydrogen atoms and their ions can be found in [75].

The meaning of the symbolic expression indicating a nuclear reaction should be as follows:

$$\begin{array}{l} \text{initial} \\ \text{nuclide} \end{array} \left(\begin{array}{l} \text{incoming particles} \quad , \quad \text{outgoing particles} \\ \text{or quanta} \qquad\qquad \text{or quanta} \end{array} \right) \begin{array}{l} \text{final} \\ \text{nuclide} \end{array}$$

Examples $^{14}\text{N}(\alpha, \text{p})^{17}\text{O}$, $\quad$ $^{59}\text{Co}(\text{n}, \gamma)^{60}\text{Co}$
$\qquad\qquad$ $^{23}\text{Na}(\gamma, 3\text{n})^{20}\text{Na}$, $\quad$ $^{31}\text{P}(\gamma, \text{pn})^{29}\text{Si}$

One can also use the standard notation from kinetics (see Section 2.12, p. 63).

Examples $^{14}_{7}\text{N} + \alpha \rightarrow {}^{17}_{8}\text{O} + \text{p}$
$\qquad\qquad$ $\text{n} \rightarrow \text{p} + \text{e} + \bar{\nu}_e$

(iii) Chemical formulae

As in the case of chemical symbols of elements, chemical formulae have two different meanings:

(a) On a *microscopic* level they denote one atom, one molecule, one ion, one radical, etc. The number of atoms in an entity (always an integer) is indicated by a right subscript, the numeral 1 being omitted. Groups of atoms may be enclosed in parentheses. Charge numbers of ions and excitation symbols are added as right superscripts to the formula. The radical nature of an entity may be expressed by adding a dot to the symbol. The nomenclature for radicals, ions, radical ions, and related species is described in [76, 77].

Examples   Xe, N_2, C_6H_6, Na^+, $SO_4{}^{2-}$

$(CH_3)_3COH$ (a 2-methyl-2-propanol molecule)

NO_2^* (an excited nitrogen dioxide molecule)

NO   (a nitrogen oxide molecule)

$NO^{\bullet}$ (a nitrogen oxide molecule, stressing its free radical character)

In writing the formula for a complex ion, spacing for charge number can be added (staggered arrangement), as well as parentheses: $SO_4{}^{2-}$, $(SO_4)^{2-}$. The staggered arrangement is now recommended [74].

Specific electronic states of entities (atoms, molecules, ions) can be denoted by giving the electronic term symbol (see Section 2.6.3, p. 32) in parentheses. Vibrational and rotational states can be specified by giving the corresponding quantum numbers (see Section 2.6, p. 25 and 33).

Examples $Hg(^3P_1)$ a mercury atom in the triplet-P-one state

$HF(v = 2, J = 6)$ a hydrogen fluoride molecule in the vibrational state $v = 2$ and the rotational state $J = 6$

$H_2O^+(^2A_1)$ a water molecule ion in the doublet-A-one state

(b) On a *macroscopic* level a formula denotes a sample of a chemical substance (not necessarily stable, or capable of existing in isolated form). The chemical composition is denoted by right subscripts (not necessarily integers; the numeral 1 being omitted). A "formula unit" (which is *not* a unit of a quantity!) is an entity specified as a group of atoms (see (iv) and (v) below).

Examples   Na, Na^+, NaCl, $Fe_{0.91}S$, $XePtF_6$, NaCl

The formula can be used in expressions like $\rho(H_2SO_4)$, mass density of sulfuric acid. When specifying amount of substance the formula is often multiplied with a factor, normally a small integer or a fraction, see examples in (iv) and (v). Less formally, the formula is often used as a shorthand ("reacting with H_2SO_4"). Chemical formulae may be written in different ways according to the information they convey [15, 74, 78, 79]:

Formula	Information conveyed	Example	Notes
empirical	stoichiometric proportion only	CH_2O	1
molecular	in accord with molecular mass	$C_3H_6O_3$	
structural	structural arrangement of atoms	$CH_3CH(OH)COOH$	1
connectivity	connectivity		2
stereochemical	stereochemical configuration		2
	Fischer projection		3
resonance structure	electronic arrangement		2, 4

(1) Molecules differing only in isotopic composition are called isotopomers or isotopologues. For example, CH_2O, CHDO, CD_2O and $CH_2{}^{17}O$ are all isotopomers or isotopologues of the formaldehyde molecule. It has been suggested [16] to reserve isotopomer for molecules of the same isotopic

(iv) Equations for chemical reactions

(a) On a *microscopic* level the reaction equation represents an elementary reaction (an event involving single atoms, molecules, and radicals), or the sum of a set of such reactions. Stoichiometric numbers are ± 1 (sometimes ± 2). A single arrow is used to connect reactants and products in an elementary reaction. An equal sign is used for the "net" reaction, the result of a set of elementary reactions (see Section 2.12.1, p. 68).

$$H + Br_2 \rightarrow HBr + Br \quad \text{one elementary step in HBr formation}$$
$$H_2 + Br_2 = 2\,HBr \quad \text{the sum of several such elementary steps}$$

(b) On a *macroscopic* level, different symbols are used connecting the reactants and products in the reaction equation, with the following meanings:

$$H_2 + Br_2 = 2\,HBr \quad \text{stoichiometric equation}$$
$$H_2 + Br_2 \rightarrow 2\,HBr \quad \text{net forward reaction}$$
$$H_2 + Br_2 \leftrightarrows 2\,HBr \quad \text{reaction, both directions}$$
$$H_2 + Br_2 \rightleftharpoons 2\,HBr \quad \text{equilibrium}$$

The two-sided arrow $\leftrightarrow$ should not be used for reactions to avoid confusion with resonance structures (see Section 2.10.1 (iii), p. 50).

Stoichiometric numbers are not unique. One and the same reaction can be expressed in different ways (without any change in meaning).

Examples The formation of hydrogen bromide from the elements can equally well be written in either of these two ways
$$(1/2)\,H_2 + (1/2)\,Br_2 = HBr$$
$$H_2 + Br_2 = 2\,HBr$$

Ions not taking part in a reaction ("spectator ions") are often removed from an equation. Redox equations are often written so that the value of the stoichiometric number for the electrons transferred (which are normally omitted from the overall equation) is equal to ± 1.

Example $(1/5)\,KMn^{VII}O_4 + (8/5)\,HCl = (1/5)\,Mn^{II}Cl_2 + (1/2)\,Cl_2 + (1/5)\,KCl + (4/5)\,H_2O$

The oxidation of chloride by permanganate in acid solution can thus be represented in several (equally correct) ways.

Examples $KMnO_4 + 8\,HCl = MnCl_2 + (5/2)\,Cl_2 + KCl + 4\,H_2O$
$MnO_4^- + 5\,Cl^- + 8\,H^+ = Mn^{2+} + (5/2)\,Cl_2 + 4\,H_2O$
$(1/5)\,MnO_4^- + Cl^- + (8/5)\,H^+ = (1/5)\,Mn^{2+} + (1/2)\,Cl_2 + (4/5)\,H_2O$

Similarly a reaction in an electrochemical cell may be written so that the electron number of an electrochemical cell reaction, z, (see Section 2.13, p. 71) is equal to one:

Example $(1/3)\,In^0(s) + (1/2)\,Hg_2^I SO_4(s) = (1/6)\,In_2^{III}(SO_4)_3(aq) + Hg^0(l)$

where the symbols in parentheses denote the state of aggregation (see Section 2.10.1 (vi), p. 54).

(Notes continued)

(1) (continued) composition but different structure, such as $CD_3CH(OH)COOH$ and $CH_3CD(OD)COOD$ for which one also uses isotope-isomer.

(2) The lines in the connectivity, stereochemical, or resonance structure formula represent single, double, or triple bonds. They are also used as representing lone-pair electrons (single valence bond).

(3) In the Fischer projection the horizontal substituents are interpreted as above the plane of the paper whereas the vertical substituents are interpreted as behind the plane of the paper.

(4) The two-sided arrow represents the arrangement of valence electrons in resonance structures, such as benzene C_6H_6 and does not signify a reaction equation (see Section 2.10.1 (iv), this page).

Symbolic Notation: A general chemical equation can be written as

$$0 = \sum_j \nu_j \, B_j$$

where B_j denotes a species in the reaction and ν_j the corresponding stoichiometric number (negative for reactants and positive for products). The ammonia synthesis is equally well expressed in these two possible ways:

(i) $(1/2)\,N_2 + (3/2)\,H_2 = NH_3$ $\nu(N_2) = -1/2$, $\nu(H_2) = -3/2$, $\nu(NH_3) = +1$
(ii) $N_2 + 3\,H_2 = 2\,NH_3$ $\nu(N_2) = -1$, $\nu(H_2) = -3$, $\nu(NH_3) = +2$

The changes $\Delta n_j = n_j - n_{j,0}$ in the amounts of any reactant and product j during the course of the reaction is governed by one parameter, the extent of reaction ξ, through the equation

$$n_j = n_{j,0} + \nu_j \, \xi$$

The extent of reaction depends on how the reaction is written, but it is independent of which entity in the reaction is used in the definition. Thus, for reaction (i), when $\xi = 2$ mol, then $\Delta n(N_2) = -1$ mol, $\Delta n(H_2) = -3$ mol and $\Delta n(NH_3) = +2$ mol. For reaction (ii), when $\Delta n(N_2) = -1$ mol then $\xi = 1$ mol.

Matrix Notation: For multi-reaction systems it is convenient to write the chemical equations in matrix form

$$A \, \nu = 0$$

where A is the conservation (or formula) matrix with elements A_{ij} representing the number of atoms of the ith element in the jth reaction species (reactant or product entity) and ν is the stoichiometric number matrix with elements ν_{jk} being the stoichiometric numbers of the jth reaction species in the kth reaction. When there are N_s reacting species involved in the system consisting of N_e elements A becomes an $N_e \times N_s$ matrix. Its nullity, $N(A) = N_s - \text{rank}(A)$, gives the number of independent chemical reactions, N_r, and the $N_s \times N_r$ stoichiometric number matrix, ν, can be determined as the null space of A. 0 is an $N_e \times N_r$ zero matrix [80].

(v) Amount of substance and the specification of entities

The quantity "amount of substance" or "chemical amount" ("Stoffmenge" in German, "quantité de matière" in French) has been used by chemists for a long time without a proper name. It was simply referred to as the "number of moles". This practice should be abandoned: the name of a physical quantity should not contain the name of a unit (few would use "number of metres" as a synonym for "length").

The amount of substance is proportional to the number of specified elementary entities of that substance; the proportionality constant is the same for all substances and is the reciprocal of the Avogadro constant. The elementary entities may be chosen as convenient, not necessarily as physically real individual particles. In the examples below, $(1/2)\,Cl_2$, $(1/5)\,KMnO_4$, etc. are artificial in the sense that no such elementary entities exists. Since the amount of substance and all physical quantities derived from it depend on this choice it is essential to specify the entities to avoid ambiguities.

Examples	n_{Cl}, $n(Cl)$	amount of Cl, amount of chlorine atoms
	$n(Cl_2)$	amount of Cl_2, amount of chlorine molecules
	$n(H_2SO_4)$	amount of (entities) H_2SO_4
	$n((1/5)\,KMnO_4)$	amount of (entities) $(1/5)\,KMnO_4$
	$M(P_4)$	molar mass of tetraphosphorus P_4
	c_{Cl^-}, $c(Cl^-)$, $[Cl^-]$	amount concentration of Cl^-
	$\rho(H_2SO_4)$	mass density of sulfuric acid
	$\Lambda(MgSO_4)$	molar conductivity of (entities of) $MgSO_4$
	$\Lambda((1/2)\,MgSO_4)$	molar conductivity of (entities of) $(1/2)\,MgSO_4$
	$\lambda(Mg^{2+})$	ionic conductivity of (entities of) Mg^{2+}
	$\lambda((1/2)\,Mg^{2+})$	ionic conductivity of (entities of) $(1/2)\,Mg^{2+}$

Using definitions of various quantities we can derive equations like

$$n((1/5)\,KMnO_4) = 5n(KMnO_4)$$
$$\lambda((1/2)\,Mg^{2+}) = (1/2)\lambda(Mg^{2+})$$
$$[(1/2)\,H_2SO_4] = 2\,[H_2SO_4]$$

(See also examples in Section 3.3, p. 88.)

Note that "amount of sulfur" is an ambiguous statement, because it might imply $n(S)$, $n(S_8)$, or $n(S_2)$, etc. In most cases analogous statements are less ambiguous. Thus for compounds the implied entity is usually the molecule or the common formula entity, and for solid metals it is the atom.

Examples	"2 mol of water" implies $n(H_2O) = 2$ mol
	"0.5 mol of sodium chloride" implies $n(NaCl) = 0.5$ mol
	"3 mmol of iron" implies $n(Fe) = 3$ mmol

Such statements should be avoided whenever there might be ambiguity.

In the equation $pV = nRT$ and in equations involving colligative properties, the entity implied in the definition of n should be an independently translating particle (a whole molecule for a gas), whose nature is unimportant. Quantities that describe compositions of mixtures can be found in [72].

(vi) States of aggregation

The following one-, two- or three-letter symbols are used to represent the states of aggregation of chemical species [1.j]. The letters are appended to the formula symbol in parentheses, and should be printed in Roman (upright) type without a full stop (period).

a, ads	species adsorbed on a surface	g	gas or vapor
am	amorphous solid	l	liquid
aq	aqueous solution	lc	liquid crystal
aq, ∞	aqueous solution at infinite dilution	mon	monomeric form
cd	condensed phase	n	nematic phase
	(i.e., solid or liquid)	pol	polymeric form
cr	crystalline	s	solid
f	fluid phase	sln	solution
	(i.e., gas or liquid)	vit	vitreous substance

Examples
         HCl(g)   hydrogen chloride in the gaseous state
         $C_V(\mathrm{f})$ heat capacity of a fluid at constant volume
         $V_\mathrm{m}(\mathrm{lc})$ molar volume of a liquid crystal
         $U(\mathrm{cr})$ internal energy of a crystalline solid
         $MnO_2(\mathrm{am})$ manganese dioxide as an amorphous solid
         $MnO_2(\mathrm{cr, I})$ manganese dioxide as crystal form I
         NaOH(aq)   aqueous solution of sodium hydroxide
         NaOH(aq, ∞)   aqueous solution of sodium hydroxide at infinite dilution
         $\Delta_\mathrm{f} H^\ominus(\mathrm{H_2O}, \mathrm{l})$ standard enthalpy of formation of liquid water

The symbols g, l, etc. to denote gas phase, liquid phase, etc., are also sometimes used as a right superscript, and the Greek letter symbols α, β, etc. may be similarly used to denote phase α, phase β, etc. in a general notation.

Examples $V_\mathrm{m}^{\ \mathrm{l}}$, $V_\mathrm{m}^{\ \mathrm{s}}$ molar volume of the liquid phase, ... of the solid phase
                   $S_\mathrm{m}^{\ \alpha}$, $S_\mathrm{m}^{\ \beta}$ molar entropy of phase α, ... of phase β

2.11 CHEMICAL THERMODYNAMICS

The names and symbols of the more generally used quantities given here are also recommended by IUPAP [4] and by ISO [5.d,5.h]. Additional information can be found in [1.d,1.j] and [81].

Name	Symbol	Definition	SI unit	Notes
heat	Q, q		J	1
work	W, w		J	1
internal energy	U	$dU = dQ + dW$	J	1
enthalpy	H	$H = U + pV$	J	
thermodynamic temperature	$T, (\Theta)$		K	
International temperature	T_{90}		K	2
Celsius temperature	θ, t	$\theta/°C = T/K - 273.15$	°C	3
entropy	S	$dS = dQ_{rev}/T$	J K^{-1}	
Helmholtz energy, (Helmholtz function)	A, F	$A = U - TS$	J	4
Gibbs energy, (Gibbs function)	G	$G = H - TS$	J	
Massieu function	J	$J = -A/T$	J K^{-1}	
Planck function	Y	$Y = -G/T$	J K^{-1}	
surface tension	γ, σ	$\gamma = (\partial G/\partial A_s)_{T,p,n_i}$	J m^{-2}, N m^{-1}	
molar quantity X	$X_m, (\overline{X})$	$X_m = X/n$	$[X]$/mol	5, 6
specific quantity X	x	$x = X/m$	$[X]$/kg	5, 6
pressure coefficient	β	$\beta = (\partial p/\partial T)_V$	Pa K^{-1}	
relative pressure coefficient	α_p	$\alpha_p = (1/p)(\partial p/\partial T)_V$	K^{-1}	
compressibility,				
isothermal	κ_T	$\kappa_T = -(1/V)(\partial V/\partial p)_T$	Pa^{-1}	
isentropic	κ_S	$\kappa_S = -(1/V)(\partial V/\partial p)_S$	Pa^{-1}	
linear expansion coefficient	α_l	$\alpha_l = (1/l)(\partial l/\partial T)$	K^{-1}	
cubic expansion coefficient	α, α_V, γ	$\alpha = (1/V)(\partial V/\partial T)_p$	K^{-1}	7
heat capacity,				
at constant pressure	C_p	$C_p = (\partial H/\partial T)_p$	J K^{-1}	
at constant volume	C_V	$C_V = (\partial U/\partial T)_V$	J K^{-1}	

(1) In the differential form, d denotes an inexact differential. The given equation in integrated form is $\Delta U = Q + W$. $Q > 0$ and $W > 0$ indicate an increase in the energy of the system.

(2) This temperature is defined by the "International Temperature Scale of 1990 (ITS-90)" for which specific definitions are prescribed [82–84]. However, the CIPM, in its 94th Meeting of October 2005, has approved a Recommendation T3 (2005) of the Comité Consultatif de Thermométrie, where the ITS-90 becomes one of several "mises en pratique" of the kelvin definition. The Technical Annex (2005) is available as Doc. CCT_05_33 [85]. It concerns the definition of a reference isotopic composition of hydrogen and water, when used for the realization of fixed points of the "mises en pratique".

(3) This quantity is sometimes misnamed "centigrade temperature".

(4) The symbol F is sometimes used as an alternate symbol for the Helmholtz energy.

(5) The definition applies to pure substance. However, the concept of molar and specific quantities (see Section 1.4, p. 6) may also be applied to mixtures. n is the amount of substance (see Section

Name	Symbol	Definition	SI unit	Notes
ratio of heat capacities	$\gamma, (\kappa)$	$\gamma = C_p/C_V$	1	
Joule-Thomson coefficient	μ, μ_{JT}	$\mu = (\partial T/\partial p)_H$	K Pa^{-1}	
thermal power	Φ, P	$\Phi = \mathrm{d}Q/\mathrm{d}t$	W	
virial coefficient,				
second	B	$pV_\mathrm{m} = RT(1 + B/V_\mathrm{m}+$	m^3 mol^{-1}	8
third	C	$C/V_\mathrm{m}{}^2 + \cdots)$	m^6 mol^{-2}	8
van der Waals	a	$(p + a/V_\mathrm{m}{}^2)(V_\mathrm{m} - b) = RT$	J m^3 mol^{-2}	9
coefficients	b		m^3 mol^{-1}	9
compression factor,	Z	$Z = pV_\mathrm{m}/RT$	1	
(compressibility factor)				
partial molar	$X_\mathrm{B}, (\overline{X}_\mathrm{B})$	$X_\mathrm{B} = (\partial X/\partial n_\mathrm{B})_{T,p,n_{j\neq B}}$	$[X]$/mol	10
quantity X				
chemical potential,	μ	$\mu_\mathrm{B} = (\partial G/\partial n_\mathrm{B})_{T,p,n_{j\neq B}}$	J mol^{-1}	11
(partial molar Gibbs				
energy)				
standard chemical	$\mu^{\ominus}, \mu^{\circ}$		J mol^{-1}	12
potential				
absolute activity	λ	$\lambda_\mathrm{B} = \exp(\mu_\mathrm{B}/RT)$	1	11
(relative) activity	a	$a_\mathrm{B} = \exp\left(\dfrac{\mu_\mathrm{B} - \mu_\mathrm{B}{}^{\ominus}}{RT}\right)$	1	11, 13
standard partial	$H_\mathrm{B}{}^{\ominus}$	$H_\mathrm{B}{}^{\ominus} = \mu_\mathrm{B}{}^{\ominus} + TS_\mathrm{B}{}^{\ominus}$	J mol^{-1}	11, 12
molar enthalpy				
standard partial	$S_\mathrm{B}{}^{\ominus}$	$S_\mathrm{B}{}^{\ominus} = -(\partial \mu_\mathrm{B}{}^{\ominus}/\partial T)_p$	J mol^{-1} K^{-1}	11, 12
molar entropy				

(Notes continued)

(5) (continued) 2.10, notes 1 and 2, p. 47).

(6) X is an extensive quantity, whose SI unit is $[X]$. In the case of molar quantities the entities should be specified.

Example $V_{\mathrm{m,B}} = V_\mathrm{m}(\mathrm{B}) = V/n_\mathrm{B}$ denotes the molar volume of B

(7) This quantity is also called the coefficient of thermal expansion, or the expansivity coefficient.

(8) Another set of "pressure virial coefficients" may be defined by

$$pV_\mathrm{m} = RT(1 + B_p\, p + C_p\, p^2 + \cdots)$$

(9) For a gas satisfying the van der Waals equation of state, given in the definition, the second virial coefficient is related to the parameters a and b in the van der Waals equation by

$$B = b - a/RT$$

(10) The symbol applies to entities B which should be specified. The bar may be used to distinguish partial molar X from X when necessary.

Example   The partial molar volume of Na_2SO_4 in aqueous solution may be denoted $\overline{V}(Na_2SO_4, \mathrm{aq})$, in order to distinguish it from the volume of the solution $V(Na_2SO_4, \mathrm{aq})$.

(11) The definition applies to entities B which should be specified. The chemical potential can be defined equivalently by the corresponding partial derivatives of other thermodynamic functions (U, H, A).

(12) The symbol $^{\ominus}$ or $^{\circ}$ is used to indicate standard. They are equally acceptable. Definitions of standard states are discussed in Section 2.11.1 (iv), p. 61. Whenever a standard chemical potential

Name	Symbol	Definition	SI unit	Notes
standard reaction Gibbs energy (function)	$\Delta_r G^{\ominus}$	$\Delta_r G^{\ominus} = \sum_B \nu_B \mu_B^{\ominus}$	J mol^{-1}	12, 14 15, 16
affinity of reaction	$A, \mathcal{A}$	$A = -(\partial G/\partial \xi)_{p,T}$ $= -\sum_B \nu_B \mu_B$	J mol^{-1}	15
standard reaction enthalpy	$\Delta_r H^{\ominus}$	$\Delta_r H^{\ominus} = \sum_B \nu_B H_B^{\ominus}$	J mol^{-1}	12, 14 15, 16
standard reaction entropy	$\Delta_r S^{\ominus}$	$\Delta_r S^{\ominus} = \sum_B \nu_B S_B^{\ominus}$	J mol^{-1} K^{-1}	12, 14, 15
reaction quotient	Q	$Q = \prod_B a_B^{\nu_B}$	1	17
equilibrium constant	$K^{\ominus}, K$	$K^{\ominus} = \exp(-\Delta_r G^{\ominus}/RT)$	1	12, 15, 18
equilibrium constant, pressure basis	K_p	$K_p = \prod_B p_B^{\nu_B}$	Pa$^{\Sigma \nu_B}$	15, 19
concentration basis	K_c	$K_c = \prod_B c_B^{\nu_B}$	(mol m^{-3})$^{\Sigma \nu_B}$	15, 19
molality basis	K_m	$K_m = \prod_B m_B^{\nu_B}$	(mol kg^{-1})$^{\Sigma \nu_B}$	15, 19
fugacity	$f, \widetilde{p}$	$f_B = \lambda_B \lim_{p \to 0} (p_B/\lambda_B)_T$	Pa	11
fugacity coefficient	ϕ	$\phi_B = f_B/p_B$	1	
Henry's law constant	k_H	$k_{H,B} = \lim_{x_B \to 0} (f_B/x_B)$ $= (\partial f_B/\partial x_B)_{x_B=0}$	Pa	11, 20

(Notes continued)

(12) (continued) $\mu^{\ominus}$ or a standard equilibrium constant $K^{\ominus}$ or other standard quantity is used, the standard state must be specified.

(13) In the defining equation given here, the pressure dependence of the activity has been neglected as is often done for condensed phases at atmospheric pressure. An equivalent definition is $a_B = \lambda_B/\lambda_B^{\ominus}$, where $\lambda_B^{\ominus} = \exp(\mu_B^{\ominus}/RT)$. The definition of $\mu^{\ominus}$ depends on the choice of the standard state (see Section 2.11.1 (iv), p. 61).

(14) The symbol r indicates reaction in general. In particular cases r can be replaced by another appropriate subscript, e.g. $\Delta_f H^{\ominus}$ denotes the standard molar enthalpy of formation; see Section 2.11.1 (i), p. 59 below for a list of subscripts. Δ_r can be interpreted as operator symbol $\Delta_r \overset{\text{def}}{=} \partial/\partial \xi$.

(15) The reaction must be specified for which this quantity applies.

(16) Reaction enthalpies (and reaction energies in general) are usually quoted in kJ mol^{-1}. In the older literature kcal mol^{-1} is also common, however, various calories exist. For the thermochemical calorie, 1 kcal = 4.184 kJ (see Section 7.2, p. 137).

(17) This quantity applies in general to a system which is not in equilibrium.

(18) This quantity is equal to the value of Q in equilibrium, when the affinity is zero. It is dimensionless and its value depends on the choice of standard state, which must be specified. ISO [5.h] and the IUPAC Thermodynamics Commission [81] recommend the symbol $K^{\ominus}$ and the name "standard equilibrium constant". Many chemists prefer the symbol K and the name "thermodynamic equilibrium constant".

(19) These quantities are not in general dimensionless. One can define in an analogous way an equilibrium constant in terms of fugacity K_f, etc. At low pressures K_p is approximately related to $K^{\ominus}$ by the equation $K^{\ominus} \approx K_p/(p^{\ominus})^{\Sigma \nu_B}$, and similarly in dilute solutions K_c is approximately related to $K^{\ominus}$ by $K^{\ominus} \approx K_c/(c^{\ominus})^{\Sigma \nu_B}$; however, the exact relations involve fugacity coefficients or activity coefficients [81].

Name	Symbol	Definition	SI unit	Notes
activity coefficient				
referenced to Raoult's law	f	$f_B = a_B/x_B$	1	11, 21
referenced to Henry's law				
molality basis	γ_m	$a_{m,B} = \gamma_{m,B} m_B/m^\ominus$	1	11, 22
concentration basis	γ_c	$a_{c,B} = \gamma_{c,B} c_B/c^\ominus$	1	11, 22
mole fraction basis	γ_x	$a_{x,B} = \gamma_{x,B} x_B$	1	10, 21
ionic strength,				
molality basis	I_m, I	$I_m = (1/2)\sum_i m_i z_i{}^2$	mol kg^{-1}	
concentration basis	I_c, I	$I_m = (1/2)\sum_i c_i z_i{}^2$	mol m^{-3}	
osmotic coefficient,				
molality basis	ϕ_m	$\phi_m = \dfrac{\mu_A{}^* - \mu_A}{RTM_A\sum_B m_B}$	1	23, 24
mole fraction basis	ϕ_x	$\phi_x = -\dfrac{\mu_B{}^* - \mu_B}{RT\ln x_B}$	1	22, 23
osmotic pressure	Π	$\Pi = -(RT/V_A)\ln a_A$	Pa	23, 25

(Notes continued)

(19) (continued) The equilibrium constant of dissolution of an electrolyte (describing the equilibrium between excess solid phase and solvated ions) is often called a solubility product, denoted K_{sol} or K_s (or $K_{sol}{}^\ominus$ or $K_s{}^\ominus$ as appropriate). In a similar way the equilibrium constant for an acid dissociation is often written K_a, for base hydrolysis K_b, and for water dissociation K_w.

(20) Henry's law is sometimes expressed in terms of molalities or concentration and then the corresponding units of Henry's law constant are Pa kg mol^{-1} or Pa m^3 mol^{-1}, respectively.

(21) This quantity applies to pure phases, substances in mixtures, or solvents.

(22) This quantity applies to solutes.

(23) A is the solvent, B is one or more solutes.

(24) The entities B are independent solute molecules, ions, etc. regardless of their nature. Their amount is sometimes expressed in osmoles (meaning a mole of osmotically active entities), but this use is discouraged.

(25) This definition of osmotic pressure applies to an incompressible fluid.

2.11.1   Other symbols and conventions in chemical thermodynamics

A more extensive description of this subject can be found in [81].

(i) Symbols used as subscripts to denote a physical chemical process or reaction
These symbols should be printed in Roman (upright) type, without a full stop (period).

adsorption	ads
atomization	at
combustion reaction	c
dilution (of a solution)	dil
displacement	dpl

formation reaction	f
immersion	imm
melting, fusion (solid $\rightarrow$ liquid)	fus
mixing of fluids	mix
reaction in general	r
solution (of solute in solvent)	sol
sublimation (solid $\rightarrow$ gas)	sub
transition (between two phases)	trs
triple point	tp
vaporization, evaporation (liquid $\rightarrow$ gas)	vap

(ii) Recommended superscripts

activated complex, transition state	$\ddagger, \neq$
apparent	app
excess quantity	E
ideal	id
infinite dilution	∞
pure substance	$*$
standard	$\ominus$, $\circ$

(iii) Examples of the use of the symbol Δ

The symbol Δ denotes a change in an extensive thermodynamic quantity for a process. The addition of a subscript to the Δ denotes a change in the property.

Examples $\Delta_{vap}H = \Delta_l^g H = H(g) - H(l)$ for the molar enthalpy of vaporization.
$\Delta_{vap}H = 40.7$ kJ mol^{-1} for water at 100 °C under its own vapor pressure.
This can also be written ΔH_{vap}, but this usage is not recommended.

The subscript r is used to denote changes associated with a *chemical reaction*. Symbols such as $\Delta_r H$ are defined by the equation

$$\Delta_r H = \sum_B \nu_B H_B = (\partial H / \partial \xi)_{T,p}$$

It is thus essential to specify the stoichiometric reaction equation when giving numerical values for such quantities in order to define the extent of reaction ξ and the value of the stoichiometric numbers ν_B .

Example $N_2(g) + 3 H_2(g) = 2 NH_3(g)$, $\Delta_r H^{\ominus}(298.15 \text{ K}) = -92.4$ kJ mol^{-1}
$\Delta_r S^{\ominus}(298.15 \text{ K}) = -199$ J mol^{-1} K^{-1}

The mol^{-1} in the units identifies the quantities in this example as the change per extent of reaction. They may be called the molar enthalpy and entropy of reaction, and a subscript m may be added to the symbol to emphasize the difference from the integral quantities if desired.

The *standard reaction quantities* are particularly important. They are defined by the equations

$$\Delta_r H^{\ominus} = \sum_B \nu_B H_B{}^{\ominus}$$

$$\Delta_r S^{\ominus} = \sum_B \nu_B S_B{}^{\ominus}$$

$$\Delta_r G^{\ominus} = \sum_B \nu_B \mu_B{}^{\ominus}$$

It is important to specify notation with care for these symbols. The relation to the affinity of the reaction is

$$-A = \Delta_r G = \Delta_r G^\ominus + RT \ln \left(\prod_B a_B^{\nu_B} \right)$$

and the relation to the standard equilibrium constant is $\Delta_r G^\ominus = -RT \ln K^\ominus$. The product of the activity coefficients is the reaction quotient Q, see p. 58.

The term *combustion* and symbol c denote the complete oxidation of a substance. For the definition of complete oxidation of substances containing elements other than C, H and O see [86]. The corresponding reaction equation is written so that the stoichiometric number ν of the substance is -1.

Example   The standard enthalpy of combustion of gaseous methane is
$\Delta_c H^\ominus(\text{CH}_4, \text{g}, 298.15\text{ K}) = -890.3\text{ kJ mol}^{-1}$, implying the reaction
$\text{CH}_4(\text{g}) + 2\,\text{O}_2(\text{g}) \rightarrow \text{CO}_2(\text{g}) + 2\,\text{H}_2\text{O}(\text{l})$.

The term *formation* and symbol f denote the formation of the substance from elements in their reference state (usually the most stable state of each element at the chosen temperature and standard pressure). The corresponding reaction equation is written so that the stoichiometric number ν of the substance is $+1$.

Example   The standard entropy of formation of crystalline mercury(II) chloride is
$\Delta_f S^\ominus(\text{HgCl}_2, \text{cr}, 298.15\text{ K}) = -154.3\text{ J mol}^{-1}\text{K}^{-1}$, implying the reaction
$\text{Hg}(\text{l}) + \text{Cl}_2(\text{g}) \rightarrow \text{HgCl}_2(\text{cr})$.

The term *atomization*, symbol at, denotes a process in which a substance is separated into its constituent atoms in the ground state in the gas phase. The corresponding reaction equation is written so that the stoichiometric number ν of the substance is -1.

Example   The standard (internal) energy of atomization of liquid water is
$\Delta_{at} U^\ominus(\text{H}_2\text{O}, \text{l}) = 625\text{ kJ mol}^{-1}$, implying the reaction
$\text{H}_2\text{O}(\text{l}) \rightarrow 2\,\text{H}(\text{g}) + \text{O}(\text{g})$.

(iv) Standard states [1.j] and [81]

The standard chemical potential of substance B at temperature $T, \mu_B^\ominus(T)$, is the value of the chemical potential under standard conditions, specified as follows. Three differently defined standard states are recognized.

For a gas phase. The standard state for a gaseous substance, whether pure or in a gaseous mixture, is the (hypothetical) state of the pure substance B in the gaseous phase at the standard pressure $p = p^\ominus$ and exhibiting ideal gas behavior. The standard chemical potential is defined as

$$\mu_B^\ominus(T) = \lim_{p \to 0} [\mu_B(T, p, y_B, ...) - RT \ln(y_B p / p^\ominus)]$$

For a pure phase, or a mixture, or a solvent, in the liquid or solid state. The standard state for a liquid or solid substance, whether pure or in a mixture, or for a solvent, is the state of the pure substance B in the liquid or solid phase at the standard pressure $p = p^\ominus$. The standard chemical potential is defined as

$$\mu_B(T) = \mu_B^*(T, p^\ominus)$$

For a solute in solution. For a solute in a liquid or solid solution the standard state is referenced to the ideal dilute behavior of the solute. It is the (hypothetical) state of solute B at the standard molality $m^\ominus$, standard pressure $p^\ominus$, and behaving like the infinitely dilute solution. The standard chemical potential is defined as

$$\mu_B^{\ominus}(T) = [\mu_B(T, p^{\ominus}, m_B, ...) - RT \ln(m_B/m^{\ominus})]^{\infty}$$

The chemical potential of the solute B as a function of the molality m_B at constant pressure $p = p^{\ominus}$ is then given by the expression

$$\mu_B(m_B) = \mu_B^{\ominus} + RT \ln(m_B \gamma_{m,B}/m^{\ominus})$$

Sometimes (amount) concentration c is used as a variable in place of molality m; both of the above equations then have c in place of m throughout. Occasionally mole fraction x is used in place of m; both of the above equations then have x in place of m throughout, and $x^{\ominus} = 1$. Although the standard state of a solute is always referenced to ideal dilute behavior, the definition of the standard state and the value of the standard chemical potential $\mu^{\ominus}$ are different depending on whether molality m, concentration c, or mole fraction x is used as a variable.

(v) Standard pressures, molality, and concentration
In principle one may choose any values for the standard pressure $p^{\ominus}$, the standard molality $m^{\ominus}$, and the standard concentration $c^{\ominus}$, although the choice must be specified. For example, in tabulating data appropriate to high pressure chemistry it may be convenient to choose a value of $p^{\ominus} = 100$ MPa ($= 1$ kbar).

In practice, however, the most common choice is

$$
\begin{aligned}
p^{\ominus} &= 0.1 \text{ MPa} = 100 \text{ kPa } (= 1 \text{ bar}) \\
m^{\ominus} &= 1 \text{ mol kg}^{-1} \\
c^{\ominus} &= 1 \text{ mol dm}^{-3}
\end{aligned}
$$

These values for $m^{\ominus}$ and $c^{\ominus}$ are universally accepted. The value for $p^{\ominus} = 100$ kPa, is the IUPAC recommendation since 1982 [1.j], and is recommended for tabulating thermodynamic data. Prior to 1982 the standard pressure was usually taken to be $p^{\ominus} = 101\,325$ Pa ($= 1$ atm, called the *standard atmosphere*). In any case, the value for $p^{\ominus}$ should be specified.

The conversion of values corresponding to different $p^{\ominus}$ is described in [87–89]. The newer value of $p^{\ominus}$, 100 kPa is sometimes called the *standard state pressure*.

(vi) Biochemical standard states
Special standard states that are close to physiological conditions are often chosen. The biochemical standard state is often chosen at $[H^+] = 10^{-7}$ mol dm^{-3}. The concentrations of the solutes may be grouped together as for example, the total phosphate concentration rather than the concentration of each component, (H_3PO_4, $H_2PO_4^-$, HPO_4^{2-}, PO_4^{3-}), separately. Standard and other reference states must be specified with care [90,91].

(vii) Thermodynamic properties
Values of many thermodynamic quantities represent basic chemical properties of substances and serve for further calculations. Extensive tabulations exist, e.g. [92–96]. Special care has to be taken in reporting the data and their uncertainties [97,98].

(viii) Reference state (of an element)
The state in which the element is stable at the chosen standard state pressure and for a given temperature [16].

2.12 CHEMICAL KINETICS AND PHOTOCHEMISTRY

The recommendations given here are based on previous IUPAC recommendations [1.c,1.k] and [17], which are not in complete agreement. Recommendations regarding photochemistry are given in [60] and for recommendations on reporting of chemical kinetics data see also [99]. A glossary of terms used in chemical kinetics has been given in [100].

Name	Symbol	Definition	SI unit	Notes
rate of change of quantity X	$\dot{X}$	$\dot{X} = \mathrm{d}X/\mathrm{d}t$	(varies)	1
rate of conversion	$\dot{\xi}$	$\dot{\xi} = \mathrm{d}\xi/\mathrm{d}t$	mol s^{-1}	2
rate of concentration change (due to chemical reaction)	r_B, v_B	$r_B = \mathrm{d}c_B/\mathrm{d}t$	mol m^{-3} s^{-1}	3, 4
rate of reaction (based on amount concentration)	v, v_c	$v = \nu_B^{-1}\mathrm{d}c_B/\mathrm{d}t$ $= \dot{\xi}/V$	mol m^{-3} s^{-1}	2, 4
rate of reaction (based on number concentration), (reaction rate)	v, v_C	$v_C = \nu_B^{-1}\mathrm{d}C_B/\mathrm{d}t$	m^3 s^{-1}	
partial order of reaction	m_B, n_B	$v = k \prod_B c_B{}^{m_B}$	1	5
overall order of reaction	m, n	$m = \sum_B m_B$	1	
rate constant, rate coefficient	$k, k(T)$	$v = k \prod_B c_B{}^{m_B}$	(m^3 mol^{-1})$^{m-1}$ s^{-1}	6

(1) E.g. rate of pressure change $\dot{p} = \mathrm{d}p/\mathrm{d}t$, for which the SI unit is Pa s^{-1}, rate of entropy change $\mathrm{d}S/\mathrm{d}t$ with SI unit J K^{-1} s^{-1}.

(2) The reaction must be specified, for which this quantity applies, by giving the stoichiometric equation.

(3) The symbol and the definition apply to entities B.

(4) Note that r_B and v can also be defined on the basis of partial pressure, number concentration, surface concentration, etc., with analogous definitions. If necessary differently defined rates of reaction can be distinguished by a subscript, e.g. $v_p = \nu_B^{-1}\mathrm{d}p_B/\mathrm{d}t$, etc. Note that the rate of reaction can only be defined for a reaction of known and time-independent stoichiometry, in terms of a specified reaction equation; also the second equation for the rate of reaction follows from the first only if the volume V is constant and c_B is uniform throughout (more generally, the definition applies to local concentrations). The derivatives must be those due to the chemical reaction considered; in open systems, such as flow systems, effects due to input and output processes must be taken into account separately, as well as transport processes in general by the equation

$$(\mathrm{d}c_B/\mathrm{d}t)_{\text{total}} = (\mathrm{d}c_B/\mathrm{d}t)_{\text{reaction}} + (\mathrm{d}c_B/\mathrm{d}t)_{\text{transport}}$$

(5) The symbol applies to reactant B. The symbol m is used to avoid confusion with n for amount of substance. The order of reaction is only defined if the particular rate law applies. m is a real number.

(6) Rate constants k and pre-exponential factors A are usually quoted in either (dm^3 mol^{-1})$^{m-1}$ s^{-1} or on a molecular scale in (cm^3)$^{m-1}$ s^{-1} or (cm^3 molecule^{-1})$^{m-1}$ s^{-1}. Note that "molecule" is not a unit, but is often included for clarity, although this does not conform to accepted usage. $k(T)$ is written to stress temperature dependence. Rate constants are frequently quoted as decadic logarithms.

Example second order reaction $k = 10^{8.2}$ cm^3 mol^{-1} s^{-1} or $\lg(k/\text{cm}^3\,\text{mol}^{-1}\,\text{s}^{-1}) = 8.2$
or alternatively $k = 10^{-12.6}$ cm^3 s^{-1} or $\lg(k/\text{cm}^3\,\text{s}^{-1}) = -12.6$

Name	Symbol	Definition	SI unit	Notes
rate constant of unimolecular reaction	$k_{uni}, k_{uni}(T, c_M)$	$v = k_{uni}c_B$	s^{-1}	7
at high pressure	k_∞	$k_{uni}(c_M \to \infty)$	s^{-1}	7
at low pressure	k_0	$v = k_0 c_M c_B$	$m^3\,mol^{-1}\,s^{-1}$	7
Boltzmann constant	k, k_B		$J\,K^{-1}$	
half life	$t_{1/2}$	$c(t_{1/2}) = c(0)/2$	s	
relaxation time, lifetime, mean life	τ	$\Delta c(\tau) = \Delta c(0)/e$	s	8
(Arrhenius) activation energy	E_A, E_a	$E_A = RT^2 d(\ln k)/dT$	$J\,mol^{-1}$	9
pre-exponential factor, frequency factor	A	$A = k\exp(E_A/RT)$	$(m^3\,mol^{-1})^{m-1}\,s^{-1}$	9, 10
hard sphere radius	r		m	
collision diameter	d_{AB}	$d_{AB} = r_A + r_B$	m	
collision cross section	σ	$\sigma = \pi d_{AB}^2$	m^2	11
mean relative speed between A and B	$\bar{c}_{AB}$	$\bar{c}_{AB} = (8kT/\pi\mu)^{1/2}$	$m\,s^{-1}$	12
collision frequency				
of A with A	$z_A(A)$	$z_A(A) = \sqrt{2}\,C_A \sigma \bar{c}$	s^{-1}	11
of A with B	$z_A(B)$	$z_A(B) = C_B \sigma \bar{c}_{AB}$	s^{-1}	11

(7) The rates of unimolecular reactions show a dependence upon the concentration c_M of a collision partner M. One writes $k_{uni}(T, c_M)$ to emphasize the temperature and "pressure" dependence. At high $c_M (\to \infty)$ the dependence vanishes. At low $c_M (\to 0)$ it gives a partial order 1 of reaction with respect to c_M. In this case one defines the second order rate constant k_0.

(8) τ is defined as the time interval in which a concentration perturbation Δc falls to $1/e$ of its initial value $\Delta c(0)$. If some initial concentration of a substance decays to zero as $t \to \infty$, as in radioactive decay, then the relaxation time is the average lifetime of that substance (radioisotope). This lifetime or decay time must be distinguished from the half life.

(9) One may use as defining equation

$$E_A = -R d(\ln k)/d(1/T)$$

The term Arrhenius activation energy is to be used only for the empirical quantity defined in the table. Other empirical equations with different "activation energies", such as

$$k(T) = A' T^n \exp(-E_a'/RT)$$

are also being used. In such expressions A', n, and E_a' are taken to be temperature independent parameters.

The term activation energy is also used for an energy threshold appearing in the electronic potential (the height of the electronic energy barrier). For this "activation energy" the symbol E_0 and the term threshold energy is preferred, but E_a is also commonly used. Furthermore, E_0 may or may not include a correction for zero point energies of reactants and the transition state.

It is thus recommended to specify in any given context exactly which activation energy is meant and to reserve (Arrhenius) activation energy only and exactly for the quantity defined in the table. E_A depends on temperature and may be written $E_A(T)$.

(10) A is dependent on temperature and may be written $A(T)$.

(11) The collision cross section σ is a constant in the hard sphere collision model, but generally it is energy dependent. One may furthermore define a temperature dependent average collision cross section (see note 16). C denotes the number concentration.

(12) μ is the reduced mass, $\mu = m_A m_B/(m_A + m_B)$.

Name	Symbol	Definition	SI unit	Notes
collision density, collision number				
of A with A	Z_{AA}	$Z_{AA} = C_A z_A(A)$	$s^{-1}\,m^{-3}$	13
of A with B	Z_{AB}	$Z_{AB} = C_A z_A(B)$	$s^{-1}\,m^{-3}$	13
collision frequency factor	z_{AB}	$z_{AB} = Z_{AB}/Lc_A c_B$	$m^3\,mol^{-1}\,s^{-1}$	13
mean free path	λ	$\lambda = \bar{c}/z_A$	m	
impact parameter	b		m	14
scattering angle	θ		1, rad	15
differential cross section	I_{ji}	$I_{ji} = d\sigma_{ji}/d\Omega$	$m^2\,sr^{-1}$	16
total cross section	σ_{ji}	$\sigma_{ji} = \int I_{ji}\,d\Omega$	m^2	16
scattering matrix	$\boldsymbol{S}$		1	17
transition probability	P_{ji}	$P_{ji} = |S_{ji}|^2$	1	16, 17
standard enthalpy of activation	$\Delta^{\ddagger}H^{\ominus}, \Delta H^{\ddagger}$	$k(T) = \dfrac{kT}{h}\exp\left(\dfrac{\Delta^{\ddagger}S^{\ominus}}{R}\right)\exp\left(\dfrac{-\Delta^{\ddagger}H^{\ominus}}{RT}\right)$	$J\,mol^{-1}$	18
volume of activation	$\Delta^{\ddagger}V^{\ominus}, \Delta V^{\ddagger}$	$\Delta^{\ddagger}V^{\ominus} = -RT(\partial(\ln k)/\partial p)_T$	$m^3\,mol^{-1}$	
standard internal energy of activation	$\Delta^{\ddagger}U^{\ominus}, \Delta U^{\ddagger}$		$J\,mol^{-1}$	18
standard entropy of activation	$\Delta^{\ddagger}S^{\ominus}, \Delta S^{\ddagger}$		$J\,mol^{-1}\,K^{-1}$	18

(13) Z_{AA} and Z_{AB} are the total number of AA or AB collisions per time and volume in a system containing only A molecules, or containing molecules of two types, A and B. Three-body collisions can be treated in a similar way.

(14) The impact parameter b characterizes an individual collision between two particles; it is defined as the distance of closest approach that would result if the particle trajectories were undeflected by the collision.

(15) $\theta = 0$ implies no deflection.

(16) In all these matrix quantities the first index refers to the final and the second to the initial channel. i and j denote reactant and product channels, respectively, and Ω denotes solid angle; $d\sigma_{ji}/d\Omega$ is equal to (scattered particle current per solid angle) divided by (incident particle current per area). Elastic scattering implies $i = j$. Both I_{ji} and σ_{ji} depend on the total energy of relative motion, and may be written $I_{ji}(E)$ and $\sigma_{ji}(E)$. General collision theory leads to an expression of a rate coefficient in terms of the energy dependent total reaction cross section

$$k_{ji}(T) = \left(\frac{8k_BT}{\pi\mu}\right)^{1/2}\int_0^{\infty}\left(\frac{E}{k_BT}\right)\sigma_{ji}(E)\exp(-E/k_BT)\left(\frac{dE}{k_BT}\right)$$

where E is the translational collision energy and μ the reduced mass (see note 12). The integral can be interpreted as the thermally averaged collision cross section $\langle\sigma_{ji}\rangle$, which is to be used in the calculation of the collision frequency under thermal conditions (see note 11).

(17) The scattering matrix $\boldsymbol{S}$ is used in quantum scattering theory [101]. $\boldsymbol{S}$ is a unitary matrix $\boldsymbol{S}\boldsymbol{S}^{\dagger} = \boldsymbol{1}$. $P_{ji} = |S_{ji}|^2$ is the probability that collision partners incident in channel i will emerge in channel j.

(18) The quantities $\Delta^{\ddagger}H^{\ominus}$, $\Delta^{\ddagger}U^{\ominus}$, $\Delta^{\ddagger}S^{\ominus}$ and $\Delta^{\ddagger}G^{\ominus}$ are used in the transition state theory of chemical reactions. They are appropriately used only in connection with elementary reactions. The relation between the rate constant k and these quantities is more generally

$$k = \kappa(k_BT/h)\exp(-\Delta^{\ddagger}G^{\ominus}/RT)$$

Name	Symbol	Definition	SI unit	Notes
standard Gibbs energy of activation	$\Delta^{\ddagger}G^{\ominus}, \Delta G^{\ddagger}$	$\Delta^{\ddagger}G^{\ominus} = \Delta^{\ddagger}H^{\ominus} - T\Delta^{\ddagger}S^{\ominus}$	J mol^{-1}	18
molecular partition function for transition state	$q^{\ddagger}, q^{\ddagger}(T)$	$k_{\infty} = \kappa \dfrac{k_{\mathrm{B}}T}{h} \dfrac{q^{\ddagger}}{q_{\mathrm{A}}} \exp(-E_0/k_{\mathrm{B}}T)$	1	18
transmission coefficient	κ, γ	$\kappa = \dfrac{k_{\infty}h}{k_{\mathrm{B}}T} \dfrac{q_{\mathrm{A}}}{q^{\ddagger}} \exp(E_0/k_{\mathrm{B}}T)$	1	18
molecular partition function per volume	$\tilde{q}$	$\tilde{q} = q/V$	m^{-3}	18
specific rate constant of unimolecular reaction at E, J	$k_{\mathrm{uni}}(E, J)$	$k_{\mathrm{uni}}(E, J) = -\dfrac{\mathrm{d}(\ln c_{\mathrm{A}}(E, J))}{\mathrm{d}t}$	s^{-1}	19
specific rate constant of bimolecular reaction at collision energy E_{t}	$k_{\mathrm{bi}}(E_{\mathrm{t}})$	$k_{\mathrm{bi}}(E_{\mathrm{t}}) = \sigma(E_{\mathrm{t}})\sqrt{2E_{\mathrm{t}}/\mu}$	m^3 s^{-1}	11
density of states	$\rho(E, J, \cdots)$	$\rho(E, J, \cdots) = \mathrm{d}N(E, J, \cdots)/\mathrm{d}E$	J^{-1}	19
number (sum) of states	$N(E), W(E)$	$N(E) = \int_0^E \rho(E')\,\mathrm{d}E'$	1	19
density of states of transition state	$\rho^{\ddagger}(E)$	$\rho^{\ddagger}(E) = \mathrm{d}N^{\ddagger}(E)/\mathrm{d}E$	J^{-1}	19
number of states of transition state	$N^{\ddagger}(E), W^{\ddagger}(E)$	$N^{\ddagger}(E) = \int_{E_0}^E \rho^{\ddagger}(E')\,\mathrm{d}E'$	1	19
number of open adiabatic reaction channels	$W(E)$	$W(E) = \sum_a \mathrm{H}(E - V_a^{\mathrm{max}})$	1	19
Michaelis constant	K_{M}	$v = \dfrac{V\,[\mathrm{S}]}{K_{\mathrm{M}} + [\mathrm{S}]}$	mol m^{-3}	20
rate (coefficient) matrix	$\boldsymbol{K}, K_{fi}$	$-\dfrac{\mathrm{d}\boldsymbol{c}}{\mathrm{d}t} = \boldsymbol{K}\,\boldsymbol{c}$	s^{-1}	21
quantum yield, photochemical yield	Φ, ϕ		1	22
fluorescence rate constant	k_{f}	$\dfrac{\mathrm{d}[h\nu]_{\mathrm{f}}}{\mathrm{d}t} = k_{\mathrm{f}}\,c^*$	s^{-1}	23, 24
natural lifetime	τ_0	$\tau_0 = 1/k_{\mathrm{f}}$	s	23
natural linewidth	$\Gamma, \Gamma_{\mathrm{f}}$	$\Gamma = \hbar k_{\mathrm{f}}$	J	23
predissociation linewidth	$\Gamma_{\mathrm{p}}, \Gamma_{\mathrm{diss}}$	$\Gamma_{\mathrm{p}} = \hbar k_{\mathrm{p}}$	J	23

(18) (continued) where k has the dimensions of a first-order rate constant. A rate constant for a reaction of order n is obtained by multiplication with $(c^{\ominus})^{1-n}$. For a bimolecular reaction of an ideal gas multiply by $V^{\ominus} = (c^{\ominus})^{-1} = (kT/p^{\ominus})$. κ is a transmission coefficient, and $\Delta^{\ddagger}G^{\ominus} = \Delta^{\ddagger}H^{\ominus} - T\,\Delta^{\ddagger}S^{\ominus}$. The standard symbol $^{\ominus}$ is sometimes omitted, and these quantities are frequently written $\Delta^{\ddagger}H, \Delta^{\ddagger}U, \Delta^{\ddagger}S$ and $\Delta^{\ddagger}G$. However, the omission of the specification of a standard state leads to ambiguity in the values of these quantities. The choice of $p^{\ominus}$ and $c^{\ominus}$ in general affects the values of $\Delta^{\ddagger}H^{\ominus}, \Delta^{\ddagger}S^{\ominus}$ and $\Delta^{\ddagger}G^{\ominus}$.

The statistical mechanical formulation of transition state theory results in the equation for k_{∞} as given in the table for a unimolecular reaction in the high pressure limit, and for a bimolecular reaction one has (often with $\kappa = 1$ assumed)

(Notes continued)

(18) (continued)

$$k_{bi} = \kappa \frac{k_B T}{h} \frac{\tilde{q}^{\ddagger}}{\tilde{q}_A \tilde{q}_B} \exp(-E_0/k_B T)$$

where $q^{\ddagger}$ is the partition function of the transition state and $\tilde{q}^{\ddagger}, \tilde{q}_A, \tilde{q}_B$ are partition functions per volume for the transition state and the reaction partners A and B. E_0 is the threshold energy for reaction, below which no reaction is assumed to occur. In transition state theory, it is the difference of the zero-point level of the transition state and the zero-point level of reactants.

(19) In the theory of unimolecular reactions, it is important to consider rate constants for an ensemble of molecules at well defined energy E (and possibly other good quantum numbers such as angular momentum J etc.) with concentration $c_A(E, J, \cdots)$. Expressions equivalent to transition state theory arise for these in the framework of quasi-equilibrium theory, Rice-Ramsperger-Kassel-Marcus (RRKM) theory of the statistical adiabatic channel model (SACM). The unimolecular rate constant is then given by the RRKM expression

$$k(E) = \frac{N^{\ddagger}(E)}{h\rho(E)}$$

$\rho(E)$ is the density of states of the molecule, $N(E)$ is the total number of states of the molecule with energy less than E,

$$N(E) = \sum_i H(E - E_i) = \int_0^E \rho(E') \, dE'$$

and $N^{\ddagger}(E)$ is the corresponding number of states of the transition state. In the framework of the statistical adiabatic channel model one has

$$k(E, J) = \kappa \frac{W(E, J)}{h\rho(E, J)}$$

V_a^{max} is the maximum of the potential of adiabatic channel a, $H(x)$ is the Heaviside function (see Section 4.2, p. 107) and J is the angular momentum quantum number [102]. There may be further constants of the motion beyond J appearing here.

(20) The Michaelis constant arises in the treatment of a mechanism of enzyme catalysis

$$E + S \underset{k_{-1}}{\overset{k_1}{\rightleftarrows}} \quad (ES) \quad \underset{k_{-2}}{\overset{k_2}{\rightleftarrows}} \quad E + P$$

where E is the enzyme, S the substrate, ES the enzyme-substrate complex and P the product. V is called limiting rate [100].

(21) In generalized first-order kinetics, the rate equation can be written as a matrix equation, with the concentration vector $c = (c_1, c_2, \cdots, c_n)^T$ and the first-order rate coefficients K_{fi} as matrix elements.

(22) The quantum yield ϕ is defined in general by the *number of defined events* divided by *number of photons absorbed* [60]. For a photochemical reaction it can be defined as

$$\phi = \left| \frac{d\xi/dt}{dn_\gamma/dt} \right|$$

which is the *rate of conversion* divided by the *rate of photon absorption*.

(23) For exponential decay by spontaneous emission (fluorescence) one has for the decay of the excited state concentration c^*

$$\frac{d[h\nu]}{dt} = -\frac{dc^*}{dt} = k_f c^*$$

The full width at half maximum Γ of a lorentzian absorption line is related to the rate constant of fluorescence. For a predissociation Γ_p is related to the predissociation rate constant k_p. However, linewidths may also have other contributions in practice.

2.12.1 Other symbols, terms, and conventions used in chemical kinetics

Additional descriptions can be found in [100].

(i) Elementary reactions

Reactions that occur at the molecular level in one step are called *elementary reactions*. It is conventional to define them as unidirectional, written with a simple arrow always from left to right. The number of relevant reactant particles on the left hand side is called the *molecularity* of the elementary reaction.

Examples	$A \rightarrow B + C$	unimolecular reaction (also monomolecular)
	$A + B \rightarrow C + D$	bimolecular reaction
	$A + B + C \rightarrow D + E$	trimolecular reaction (also termolecular)

(ii) Composite mechanisms

A reaction that involves more than one elementary reaction is said to occur by a composite mechanism. The terms complex mechanism, indirect mechanism, and stepwise mechanism are also commonly used. Special types of mechanisms include chain-reaction mechanisms, catalytic reaction mechanisms, etc.

Examples A simple mechanism is composed of forward and reverse reactions

$$A \rightarrow B + C$$
$$B + C \rightarrow A$$

It is in this particular case conventional to write these in one line

$$A \rightleftarrows B + C$$

However, it is useful in kinetics to distinguish this from a net reaction, which is written either with two one-sided arrows or an "equal" sign

$$A \rightleftharpoons B + C$$
$$A = B + C$$

When one combines a composite mechanism to obtain a net reaction, one should not use the simple arrow in the resulting equation.

Example $A \rightarrow B + C$ unimolecular elementary reaction
$B + C \rightarrow D + E$ bimolecular elementary reaction

_____

$A = D + E$ net reaction (no elementary reaction, no molecularity)

It is furthermore useful to distinguish the stoichiometric equation defining the reaction rate and rate constant from the equation defining the elementary reaction and rate law.

Example Recombination of methyl radicals in the high pressure limit
The elementary reaction is $CH_3 + CH_3 \rightarrow C_2H_6$
This has a second order rate law. If one uses the stoichiometric equation
$2\,CH_3 = C_2H_6$
the definition of the reaction rate gives

$$v_C = -\frac{1}{2}\frac{dC_{CH_3}}{dt} = k\,C_{CH_3}{}^2$$

(Notes continued)

(24) The einstein is sometimes either used as the unit of amount of substance of photons, where one einstein corresponds to 1 mol of photons, or as the unit of energy where one einstein corresponds to the energy $Lh\nu$ of 1 mol of monochromatic photons with frequency ν.

(continued)   If one uses the stoichiometric equation
$$CH_3 = (1/2) \, C_2H_6$$
the definition of the reaction rate gives
$$v_C = -\frac{dC_{CH_3}}{dt} = k' \, C_{CH_3}{}^2$$
Hence $k' = 2k$.

Similar to reaction enthalpies, rate coefficients are only defined with a given stoichiometric equation. It is recommended always to state explicitly the stoichiometric equation and the differential equation for the reaction rate to avoid ambiguities. If no other statement is made, the stoichiometric coefficients should be as in the reaction equation (for the elementary reaction). However, it is not always possible to use a reaction equation as stoichiometric equation.

Example   The trimolecular reaction for hydrogen atoms
The elementary reaction is $H + H + H \rightarrow H_2 + H$
This has a third order rate law. The stoichiometric equation is
$$2 \, H = H_2$$
and the definition of the reaction rate gives
$$v_c = -\frac{1}{2}\frac{d[H]}{dt} = k \, [H]^3$$

The bimolecular reaction $H + H \rightarrow H_2$ with the same stoichiometric equation is a *different* elementary reaction (a very unlikely one), not the same as the trimolecular reaction. Other examples include catalyzed reactions such as $X + A \rightarrow X + B$ with the stoichiometric equation $A = B$ and the rate $v_c = -d[A]/dt = k[A][X] \, (\neq d[X]/dt = 0)$. Again, the unimolecular reaction $A \rightarrow B$ exists, but would be a *different* elementary reaction with the same stoichiometry.

2.13 ELECTROCHEMISTRY

Electrochemical concepts, terminology and symbols are more extensively described in [1.i] and [103–107]. For the field of electrochemical engineering see [108], for semiconductor electrochemistry and photo-electrochemical energy conversion see [109], for corrosion nomenclature see [110], for impedances in electrochemical systems see [111], for electrokinetic phenomena see [112], for electrochemical biosensors see [113], and for chemically modified electrodes (CME) see [114].

Name	Symbol	Definition	SI unit	Notes
elementary charge	e	proton charge	C	
Faraday constant	F	$F = eN_A$	C mol^{-1}	
charge number of an ion	z	$z_B = Q_B/e$	1	1
ionic strength,				
$\quad$ molality basis	I_m, I	$I_m = (1/2) \sum_i m_i z_i^2$	mol kg^{-1}	
$\quad$ concentration basis	I_c, I	$I_c = (1/2) \sum_i c_i z_i^2$	mol m^{-3}	2
mean ionic molality	$m_\pm$	$m_\pm^{(\nu_+ + \nu_-)} = m_+^{\nu_+} m_-^{\nu_-}$	mol kg^{-1}	3
mean ionic activity coefficient	$\gamma_\pm$	$\gamma_\pm^{(\nu_+ + \nu_-)} = \gamma_+^{\nu_+} \gamma_-^{\nu_-}$	1	3
mean ionic activity	$a_\pm$	$a_\pm = m_\pm \gamma_\pm / m^\ominus$	1	3, 4, 5
activity of an electrolyte	$a(A_{\nu_+} B_{\nu_-})$	$a(A_{\nu_+} B_{\nu_-}) = a_\pm^{(\nu_+ + \nu_-)}$	1	3
pH	pH	$pH = -\lg(a_{H^+})$	1	5, 6
outer electric potential	ψ		V	7
surface electric potential	χ		V	8
inner electric potential	ϕ	$\phi = \chi + \psi$	V	
Volta potential difference	$\Delta\psi$	$\Delta_\alpha^\beta \psi = \psi^\beta - \psi^\alpha$	V	9

(1) The definition applies to entities B.

(2) To avoid confusion with the cathodic current, symbol I_c (note roman subscript), the symbol I or sometimes μ (when the current is denoted by I) is used for ionic strength based on concentration.

(3) ν_+ and ν_- are the numbers of cations and anions per formula unit of an electrolyte $A_{\nu_+} B_{\nu_-}$.

$\quad$ *Example* For $Al_2(SO_4)_3$, $\nu_+ = 2$ and $\nu_- = 3$.

m_+ and m_-, and γ_+ and γ_-, are the molalities and activity coefficients of cations and anions. If the molality of $A_{\nu+} B_{\nu-}$ is m, then $m_+ = \nu_+ m$ and $m_- = \nu_- m$. A similar definition is used on a concentration scale for the mean ionic concentration $c_\pm$.

(4) $m^\ominus = 1$ mol kg^{-1}.

(5) For an individual ion, neither the activity a_+, a_- nor the activity coefficient γ_+, γ_- is experimentally measurable.

(6) The definition of pH is discussed in Section 2.13.1 (viii), p. 75. The symbol pH is an exception to the general rules for the symbols of physical quantities (Section 1.3.1, p. 5) in that it is a two-letter symbol and it is always printed in Roman (upright) type.

(7) ψ^β is the electrostatic potential of phase β due to the electric charge Q of the phase. It may be calculated from classical electrostatics. For example, for a conducting sphere with excess charge Q and radius r, placed in vacuo, $\psi = Q/4\pi\varepsilon_0 r$.

(8) The surface potential is defined as the work (divided by the charge) necessary to transfer an ideal (i.e. massless) positive charge through a dipole layer at the interphase between phase α and β. The absolute value of χ cannot be determined, only differences are measurable.

(9) $\Delta\psi$ is the outer potential difference due to the charge on phases α and β. It is a measurable quantity.

Name	Symbol	Definition	SI unit	Notes
Galvani potential difference	$\Delta\phi$	$\Delta_\alpha^\beta\phi = \phi^\beta - \phi^\alpha$	V	10
electrochemical potential	$\tilde{\mu}_B{}^\alpha$	$\tilde{\mu}_B{}^\alpha = (\partial G/\partial n_B{}^\alpha)_{T,p,n_{j\neq B}} =$ $\mu_B{}^{\alpha\ominus} + RT\ln a_B + z_B F\phi^\alpha$	J mol^{-1}	1, 11
surface charge density	σ	$\sigma = Q_s/A$	C m^{-2}	12
electrode potential	E, U		V	13
potential difference of an electrochemical cell, cell potential	E_{cell}, U_{cell}	$E_{cell} = E_R - E_L$	V	14
electron number of an electrochemical reaction (charge number)	z, n	$z = |\nu_e|$	1	15
standard electrode potential	$E^\ominus$	$E^\ominus = -\Delta_r G^\ominus/zF$	V	16, 17

(10) $\Delta\phi$ is the inner potential difference between points within the bulk phases α and β; it is measurable only if the phases are of identical composition.

(11) The electrochemical potential of ion B in a phase is the partial molar Gibbs energy of the ion. The special name "electrochemical potential", and the tilde on the symbol $\tilde{\mu}_B$, are used as a reminder that for an ion the partial molar Gibbs energy depends on the inner potential ϕ as well as on the chemical composition. The difference of electrochemical potential of ion B between two phases α and β of different chemical composition and different inner potential is given by

$$\tilde{\mu}_B{}^\alpha - \tilde{\mu}_B{}^\beta = \mu_B{}^{\alpha\ominus} - \mu_B{}^{\beta\ominus} + RT\ln\left(a_B{}^\alpha/a_B{}^\beta\right) + z_B F\left(\phi^\alpha - \phi^\beta\right)$$

It is not generally possible to separate the term depending on the composition from the term depending on the inner potential, because it is not possible to measure either the relative activity a_B of an ion, or the inner potential difference between two phases of different chemical composition. The electrochemical potential, however, is always well defined, and in a multiphase system with phases at different inner potentials, equilibrium with regard to the transfer of ion B between phases is achieved when its electrochemical potential is the same in all phases.

(12) Q_s is the charge on one side of the interface, A is the surface area. For the case of metal and semiconductor electrodes, by convention the charge refers to the electrode side.

(13) The absolute value of the electrode potential cannot be measured, so E is always reported relative to the potential of some reference electrode, e.g. that of a standard hydrogen electrode (SHE) (see Section 2.13.1 (vi), p. 74). The concept of an *absolute* electrode potential is discussed in [106].

(14) The two electrodes of an electrochemical cell can be distinguished by the subscripts L (left) and R (right), or 1 and 2. This facilitates the representation of the cell in a so-called cell diagram (see Section 2.13.1 (iii), p. 73). E_L and E_R are the electrode potentials of these two electrodes. In electrochemical engineering, the cell voltage is exclusively denoted by U. The limiting value of E_{cell} for zero current flowing through the cell, all local charge transfer and chemical equilibria being established, was formerly called emf (electromotive force). The name electromotive force and the symbol emf are no longer recommended, since a potential difference is not a force.

(15) z (or n) is the number of electrons in the balanced electrode reaction as written. It is a positive integer, equal to $|\nu_e|$, when ν_e is the stoichiometric number of the electron in the electrode reaction. n is commonly used where there is no risk of confusion with amount of substance.

(16) The symbols $\ominus$ and $\circ$ are both used to indicate standard state; they are equally acceptable.

(17) The standard potential is the value of the *equilibrium potential* of an electrode under standard conditions. $\Delta_r G^\ominus$ is the standard Gibbs energy of this electrode reaction, written as a reduction, with respect to that of the standard hydrogen electrode (see also Section 2.13.1 (vi), p. 74). The electrode potential attains its *equilibrium value* if no current flows through the cell, and all local charge

Name	Symbol	Definition	SI unit	Notes
equilibrium electrode potential (of an electro-chemical reaction)	E_{eq}	$E_{eq} = E^{\ominus} - (RT/zF)\sum_i \nu_i \ln a_i$	V	15–18
formal potential	$E^{\ominus\prime}$	$E_{eq} = E^{\ominus\prime} - (RT/zF)\sum_i \nu_i \ln(c_i/c^{\ominus})$	V	15, 19
liquid junction potential	E_j			20
electric current	I	$I = \mathrm{d}Q/\mathrm{d}t$	A	21
electric current density	$j, \boldsymbol{j}$	$j = I/A$	A m^{-2}	22
faradaic current	I_F	$I_F = I_c + I_a$	A	23
reduction rate constant	k_c	$I_c = -nFAk_c \prod_B (c_B')^{n_B}$	(varies)	1, 15, 24
oxidation rate constant	k_a	$I_a = nFAk_a \prod_B (c_B')^{n_B}$	(varies)	1, 15, 24
transfer coefficient	α, α_c	$\alpha_c = -(RT/nF)\mathrm{d}(\ln k_c)/\mathrm{d}E$	1	15, 25
overpotential	η, E_η	$\eta = E - E_{eq}$	V	
Tafel slope	b	$b = (\partial E/\partial \ln|I_F|)_{c_i,T,p}$	V	26
mass transfer coefficient	k_d	$k_{d,B} = |\nu_B| I_{\lim,B}/nFcA$	m s^{-1}	1, 15, 27

(17) (continued) transfer equilibria across phase boundaries that are represented in the cell diagram (except at possible electrolyte-electrolyte junctions) and local chemical equilibria within the phases are established.

(18) This is the Nernst equation. $\sum \nu_i \ln a_i$ refers to the electrode reaction, where a_i are the activities of the species taking part in it; ν_i are the stoichiometric numbers of these species in the equation for the electrode reaction, written as a reduction, ν_i is positive for products and negative for reactants. The equilibrium potential is also called *Nernst potential*, or *reversible potential*.

(19) It is $E^{\ominus\prime}$ which is calculated in electrochemical experiments when the concentrations of the various species are known, but not their activities. Its value depends on the composition of the electrolyte solution. The argument of ln is dimensionless, the concentration c_i is normalized through division by the standard concentration $c^{\ominus}$, usually $c^{\ominus} = 1$ mol dm^{-3} for soluble species.

(20) E_j is the Galvani potential difference between two electrolyte solutions in contact.

(21) Q is the charge transferred through the external circuit of the cell.

(22) Formally, the current density is a vector, $\mathrm{d}I = \boldsymbol{j} \cdot \boldsymbol{e}_n \mathrm{d}A$ (see Section 2.3, note 2, p. 16).

(23) I_F is the current through the electrode|solution interface, resulting from the charge transfer due to an electrode reaction proceeding as reactants $+ ne^- \rightarrow$ products. I_c is the *cathodic* partial current due to the reduction reaction, I_a is the *anodic* partial current due to the oxidation reaction. By definition I_c is negative and I_a positive. At the equilibrium potential, $I_a = -I_c = I_0$ (the exchange current) and $j_a = -j_c = j_0$ (the exchange current density). I or j may achieve limiting values, indicated by the subscript lim, in addition to the subscripts F, c, or a.

(24) For a first-order reaction the SI unit is m s^{-1}. Here n (or z) is the number of electrons transferred in the electrochemical reaction, c_B' is the concentration at the interphase, n_B is the order of the reaction with respect to entity B. The formerly used symbols k_{red}, k_f and $\overrightarrow{k}$ (for k_c) and k_{ox}, k_b and $\overleftarrow{k}$ (for k_a) are not recommended.

(25) α or α_c is also named *cathodic transfer coefficient*. Analogously the *anodic transfer coefficient* is defined as $\alpha_a = (RT/nF)\mathrm{d}(\ln k_a)/\mathrm{d}E$. At the same potential $\alpha_a = 1 - \alpha_c$. The *symmetry factor*, β, is the cathodic transfer coefficient, α_i of an elementary reaction step i, in which only one electron is transferred.

(26) The Tafel slope b is an experimental quantity from which kinetic information can be derived.

(27) The mass transfer coefficient is the flux divided by the concentration. For steady-state mass transfer, $k_{d,B} = D_B/\delta_B$ where δ_B is the diffusion layer thickness (which is model dependent) and D_B is the diffusion coefficient of entity B. For more information see [17].

Name	Symbol	Definition	SI unit	Notes
electrokinetic potential (ζ-potential)	ζ		V	
conductivity	$\kappa, (\sigma)$	$\boldsymbol{j} = \kappa \boldsymbol{E}$	S m^{-1}	22, 28
electric mobility	$u, (m)$	$u_B = |v_B|/|\boldsymbol{E}|$	m^2 V^{-1} s^{-1}	1, 29
ionic conductivity, molar conductivity of an ion	λ	$\lambda_B = |z_B|F u_B$	S m^2 mol^{-1}	1, 30, 31
molar conductivity	Λ	$\Lambda(A_{\nu_+}B_{\nu_-}) = \nu_+\lambda_+ + \nu_-\lambda_-$	S m^2 mol^{-1}	1, 30, 31
transport number	t	$t_B = \lambda_B c_B / \sum_i \lambda_i c_i = j_B / \sum_i j_i$	1	1

(28) Conductivity was formerly called specific conductance. $\boldsymbol{E}$ is the electric field strength vector.
(29) v_B is the migration velocity of entities B and $|\boldsymbol{E}|$ is the electric field strength within the phase concerned.
(30) The unit S cm^2 mol^{-1} is often used for molar conductivity. The conductivity is equal to $\kappa = \sum \lambda_i c_i$.
(31) It is important to specify the entity to which molar conductivity refers. It is standard practice to choose the entity to be $1/z_B$ of an ion of charge number z_B, in order to normalize ion charge, so that for example molar conductivities for potassium, barium and lanthanum ions would be quoted as $\lambda(K^+)$, $\lambda((1/2) Ba^{2+})$, or $\lambda((1/3) La^{3+})$ the so-called molecular conductivity of *equivalent entities* (formerly called *equivalent conductivity*) [115].

2.13.1 Sign and notation conventions in electrochemistry [1]

(i) Electrochemical cells
Electrochemical cells consist of at least two (usually metallic) electron conductors in contact with ionic conductors (electrolytes). The current flow through electrochemical cells may be zero or non-zero. Electrochemical cells with current flow can operate either as *galvanic cells*, in which chemical reactions occur spontaneously and chemical energy is converted into electrical energy, or as *electrolytic cells*, in which electrical energy is converted into chemical energy. In both cases part of the energy becomes converted into (positive or negative) heat.

(ii) Electrode
There are at present two usages for the term electrode, namely either (i) the electron conductor (usually a metal) connected to the external circuits or (ii) the half cell consisting of one electron conductor and at least one ionic conductor. The latter version has usually been favored in electrochemistry.

(iii) Representation of electrochemical cells
Electrochemical cells are represented by diagrams such as those in the following examples:

Examples   Pt(s) | H$_2$(g) | HCl(aq) | AgCl(s) | Ag(s)

Cu(s) | CuSO$_4$(aq) ¦ ZnSO$_4$(aq) | Zn(s)

Cu(s) | CuSO$_4$(aq) ⫤ KCl(aq, sat) ⫤ ZnSO$_4$(aq) | Zn(s)

A single vertical bar (|) should be used to represent a phase boundary, a dashed vertical bar (¦) to represent a junction between miscible liquids, and double dashed vertical bars (⫤) to represent a liquid junction, in which the liquid junction potential is assumed to be eliminated.

---

[1] These are in accordance with the "Stockholm Convention" of 1953 [103].

(iv) Potential difference of an electrochemical cell

The potential difference of an electrochemical cell is measured between a metallic conductor attached to the right-hand electrode of the cell diagram and an identical metallic conductor attached to the left-hand electrode. Electric potential differences can be measured only between two pieces of material of the same composition. In practice, these are almost always two pieces of copper, attached to the electrodes of the cell.

At a boundary copper | electrode material there is a contact potential difference; its value is incorporated in the constant to which the electrode potential is referred.

Due to the different rates of diffusion of anions and cations from one solution to the other, liquid junction potentials, E_j, appear whenever two solutions of different composition are immiscible or are separated by a diaphragm or some other means to avoid mixing. If equilibrium is not established at these junctions, the cells may include unknown junction potentials. Salt bridges (see, for example, the third cell in the previous Section 2.13.1 (iii)) are commonly employed to minimize or to stabilize the contributions of (liquid) junction potentials at the interface of two miscible electrolyte solutions to the measured potential difference. At the interface of two immiscible electrolyte solutions, a thermodynamic distribution potential can be established due to equilibrium charge partition.

(v) Standard potential of an electrochemical cell reaction

If no current flows through the cell, and all local charge transfer and local chemical equilibria of each electrode reaction are established, the potential difference of the cell is related to the Gibbs energy of the overall cell reaction by the equation

$$\Delta_r G = -zFE_{cell,eq}$$

assuming that junction potentials are negligible. If the reaction of the electrochemical cell as written takes place spontaneously, $\Delta_r G$ is negative and E_{cell} positive. If the cell is written with the electrodes in the inverse order $\Delta_r G$ is positive and E_{cell} is negative. The equilibrium potential of the cell, i.e. when no current flows, is given by

$$E_{cell,eq} = E^{\ominus}{}_{cell,eq} - \frac{RT}{zF} \sum_i \nu_i \ln a_i$$

a_i are the activities of the species taking part in the cell reaction and ν_i are the stoichiometric numbers of these species in the equation written for the cell reaction (see also notes 15 and 16, p. 71).

(vi) Standard electrode potential (of an electrochemical reaction)

The *standard potential* of an electrochemical reaction, abbreviated as standard potential, is defined as the standard potential of a hypothetical cell, in which the electrode (half-cell) at the left of the cell diagram is the *standard hydrogen electrode* (SHE) and the electrode at the right is the electrode in question. This implies that the cell reaction always involves the oxidation of molecular hydrogen. The standard hydrogen electrode consists of a platinum electrode in contact with a solution of H^+ at unit activity and saturated with H_2 gas with a fugacity referred to the standard pressure of 10^5 Pa (see Section 2.11.1 (v), p. 62). For a metallic electrode in equilibrium with solvated ions the cell diagram is

$$Pt \mid H_2 \mid H^+ \; \vdots \; M^{z+} \mid M$$

and relates to the reaction

$$M^{z+} + (z/2) H_2(g) = M + z H^+$$

This diagram may be abbreviated $E(M^{z+}/M)$, but the order of these symbols should not be reversed. Note that the standard hydrogen electrode as defined is limited to aqueous solutions. For more information on measuring electrode potentials in aqueous and non-aqueous systems, see [103, 107].

74

(vii) Anode, cathode

The terms anode and cathode may only be applied to electrochemical cells through which a net current flows. In a cell at equilibrium the terms plus pole and minus pole are used. An *anode* is an electrode at which the predominating electrochemical reaction is an oxidation; electrons are produced (in a galvanic cell) or extracted (in an electrolytic cell). A *cathode* is an electrode at which the predominating electrochemical reaction is a reduction which consumes the electrons from the anode, and that reach the cathode through the external circuit. Since electrons flow from lower to higher (more positive) electric potentials, in a galvanic cell they flow from the anode (the negative electrode) to the cathode (the positive electrode), while in an electrolytic cell the electrons extracted from the anode (the positive electrode) by the external source flow to the cathode (the negative electrode).

Note that in rechargeable batteries such as the lead-acid battery (lead accumulator), the positive electrode is a cathode during discharge and an anode during charge, and the negative electrode is the anode during discharge and the cathode during charge. In order to avoid confusion, it is recommended that in rechargeable batteries only the terms positive and negative electrodes (or plates) be used. For more detailed information on electrode potentials, see [109].

(viii) Definition of pH [116]

The quantity pH is defined in terms of the activity of hydrogen(1+) ions (hydrogen ions) in solution:

$$pH = pa_{H^+} = -\lg(a_{H^+}) = -\lg(m_{H^+}\gamma_{m,\,H^+}/m^{\ominus})$$

where a_{H^+} is the activity of the hydrogen(1+) (hydrogen ion) in solution, $H^+(aq)$, and $\gamma_{m,\,H^+}$ is the activity coefficient of $H^+(aq)$ on the molality basis at molality m_{H^+}. The symbol p is interpreted as an operator ($px = -\lg x$) [116, 117] (see Section 4.1, p. 103) with the unique exception of the symbol pH [117]. The symbol pH is also an exception to the rules for symbols and quantities (see Section 1.3.1, p. 5). The standard molality $m^{\ominus}$ is chosen to be equal to 1 mol kg^{-1}. Since pH is defined in terms of a quantity that cannot be measured independently, the above equation can be only regarded as a *notional* definition.

The establishment of primary pH standards requires the application of the concept of the "primary method of measurement" [116], assuring full traceability of the results of all measurements and their uncertainties. Any limitation in the theory or determination of experimental variables must be included in the estimated uncertainty of the method.

The primary method for the measurement of pH involves the use of a cell without transference, known as the *Harned cell* [118]:

$$Pt(s) \mid H_2(g) \mid Buffer\ S,\ Cl^-(aq) \mid AgCl(s) \mid Ag(s)$$

Application of the Nernst equation to the above leads to the relationship

$$E = E^{\ominus} - \frac{RT \ln 10}{F}\ \lg[(m_{H^+}\gamma_{H^+}/m^{\ominus})(m_{Cl^-}\gamma_{Cl^-}/m^{\ominus})]$$

where E is the potential difference of the cell and $E^{\ominus}$ the standard potential of the AgCl|Ag electrode. This equation can be rearranged to yield

$$-\lg(a_{H^+}\gamma_{Cl^-}) = \frac{(E-E^{\ominus})}{(RT \ln 10)/F} + \lg(m_{Cl^-}/m^{\ominus})$$

Measurements of E are made and the quantity $-\lg(a_{H^+}\gamma_{Cl^-})$ is obtained by extrapolation to $m_{Cl^-}/m^{\ominus} = 0$. The value of γ_{Cl^-} is calculated using the Bates-Guggenheim convention [117] based on Debye-Hückel theory. Then $-\lg(a_{H^+})$ is calculated and identified as pH(PS), where PS signifies primary standard. The uncertainties in the two estimates are typically ± 0.001 in $-\lg(a_{H^+}\gamma_{Cl^-})^{\ominus}$, and ± 0.003 in pH, respectively.

Materials for primary standard pH buffers must also meet the appropriate requirements for reference materials including chemical purity and stability, and applicability of the Bates-Guggenheim convention to the estimation of $\lg(\gamma_{Cl^-})$. This convention requires that the ionic strength must be $\leqslant 0.1$ mol kg^{-1}. Primary standard buffers should also lead to small liquid junction potentials when used in cells with liquid junctions. Secondary standards, pH(SS), are also available, but carry a greater uncertainty in measured values.

Practical pH measurements generally use cells involving liquid junctions in which, consequently, liquid junction potentials, E_j, are present [116]. Measurements of pH are not normally performed using the Pt|H$_2$ electrode, but rather the glass (or other H$^+$-selective) electrode, whose response factor (dE/dpH) usually deviates from the Nernst slope. The associated uncertainties are significantly larger than those associated with the fundamental measurements using the Harned cell. Nonetheless, incorporation of the uncertainties for the primary method, and for all subsequent measurements, permits the uncertainties for all procedures to be linked to the primary standards by an unbroken chain of comparisons.

Reference values for standards in D$_2$O and aqueous-organic solvent mixtures exist [119].

2.14 COLLOID AND SURFACE CHEMISTRY

The recommendations given here are based on more extensive IUPAC recommendations [1.e–1.h] and [120–123]. Catalyst characterization is described in [124] and quantities related to macromolecules in [125].

Name	Symbol	Definition	SI unit	Notes
specific surface area	a_s, a, s	$a_s = A/m$	$m^2 \ kg^{-1}$	1
surface amount of B	n_B^s		mol	2
adsorbed amount of B	n_B^a		mol	2
surface excess amount of B	n_B^σ		mol	3
surface excess concentration of B	$\Gamma_B, (\Gamma_B^\sigma)$	$\Gamma_B = n_B^\sigma/A$	$mol \ m^{-2}$	3
total surface excess concentration	$\Gamma, (\Gamma^\sigma)$	$\Gamma = \sum_i \Gamma_i$	$mol \ m^{-2}$	
area per molecule	a, σ	$a_B = A/N_B^a$	m^2	4
area per molecule in a filled monolayer	a_m, σ_m	$a_{m,B} = A/N_{m,B}$	m^2	4
surface coverage	θ	$\theta = N_B^a/N_{m,B}$	1	4
contact angle	θ		rad, 1	
film thickness	t, h, δ		m	
thickness of (surface or interfacial) layer	τ, δ, t		m	
surface tension, interfacial tension	γ, σ	$\gamma = (\partial G/\partial A_s)_{T,p,n_i}$	$N \ m^{-1}, J \ m^{-2}$	
film tension	Σ_f	$\Sigma_f = (\partial G/\partial A_f)_{T,p,n_i}$	$N \ m^{-1}$	5
Debye length of the diffuse layer	L_D	$L_D = \kappa^{-1}$	m	6
average molar masses				
number-average	M_n	$M_n = \sum_i n_i M_i / \sum_i n_i$	$kg \ mol^{-1}$	
mass-average	M_m, M_w	$M_m = \sum_i n_i M_i^2 / \sum_i n_i M_i$	$kg \ mol^{-1}$	
z-average	M_z	$M_z = \sum_i n_i M_i^3 / \sum_i n_i M_i^2$	$kg \ mol^{-1}$	
sedimentation coefficient	s	$s = v/a$	s	7
van der Waals constant	λ		J	
retarded van der Waals constant	β, B		J	
van der Waals-Hamaker constant	A_H		J	
surface pressure	π	$\pi = \gamma^0 - \gamma$	$N \ m^{-1}$	8

(1) The subscript s designates any surface area where absorption or deposition of species may occur. m designates the mass of a solid absorbent.

(2) The value of n_B^s depends on the thickness assigned to the surface layer (see also note 1, p. 47).

(3) The values of n_B^σ and Γ_B depend on the convention used to define the position of the Gibbs surface. They are given by the excess amount of B or surface concentration of B over values that would apply if each of the two bulk phases were homogeneous right up to the Gibbs dividing surface.

Additional recommendations

The superscript s denotes the properties of a surface or interfacial layer. In the presence of adsorption it may be replaced by the superscript a.

Examples	Helmholtz energy of interfacial layer	A^s
	amount of adsorbed substance	n^a
	amount of adsorbed O_2	$n^a(O_2)$ or $n(O_2, a)$
	area per molecule B in a monolayer	$a(B), a_B$

The superscript σ is used to denote a surface excess property relative to the Gibbs surface.

Example	surface excess amount	$n_B{}^\sigma$
	(or Gibbs surface amount) of B	

In general the values of Γ_A and Γ_B depend on the position chosen for the Gibbs dividing surface. However, two quantities, $\Gamma_B{}^{(A)}$ and $\Gamma_B{}^{(n)}$ (and correspondingly $n_B{}^{\sigma(A)}$ and $n_B{}^{\sigma(n)}$), may be defined in a way that is invariant to this choice (see [1.e]). $\Gamma_B{}^{(A)}$ is called the *relative* surface excess concentration of B with respect to A, or more simply the relative adsorption of B; it is the value of Γ_B when the surface is chosen to make $\Gamma_A = 0$. $\Gamma_B{}^{(n)}$ is called the *reduced* surface excess concentration of B, or more simply the reduced adsorption of B; it is the value of Γ_B when the surface is chosen to make the total excess $\Gamma = \sum_i \Gamma_i = 0$.

Properties of phases (α, β, γ) may be denoted by corresponding superscript indices.

Examples	surface tension of phase α	γ^α
	interfacial tension between phases α and β	$\gamma^{\alpha\beta}$

Symbols of thermodynamic quantities divided by surface area are usually the corresponding lower case letters; an alternative is to use a circumflex.

Example	interfacial entropy per area	$s^s (= \hat{s}^s) = S^s/A$

The following abbreviations are used in colloid chemistry:

ccc	critical coagulation concentration
cmc	critical micellisation concentration
iep	isoelectric point
pzc	point of zero charge

(Notes continued)

(3) (continued) See [1.e], and also additional recommendations on this page.

(4) $N_B{}^a$ is the number of adsorbed molecules ($N_B{}^a = Ln_B{}^a$), and $N_{m,B}{}^a$ is the number of adsorbed molecules in a filled monolayer. The definition applies to entities B.

(5) In the equation, A_f is the area of the film.

(6) The characteristic Debye length [1.e] and [123] or Debye screening length [123] L_D appears in Gouy-Chapman theory and in the theory of semiconductor space-charge.

(7) In the definition, v is the velocity of sedimentation and a is the acceleration of free fall or centrifugation. The symbol for a limiting sedimentation coefficient is $[s]$, for a reduced sedimentation coefficient s°, and for a reduced limiting sedimentation coefficient $[s^\circ]$; see [1.e] for further details.

(8) In the definition, γ^0 is the surface tension of the clean surface and γ that of the covered surface.

2.14.1 Surface structure [126]

(i) Single-crystal surface and vector designations

3D Miller indices (hkl) are used to specify a surface termination with a (hkl) bulk plane. Families of symmetry-equivalent surfaces are designated $\{hkl\}$. Vector directions are designated $[hkl]$ and families of symmetry-equivalent vector directions $< hkl >$.

(ii) Notations for stepped surfaces (for ideally terminated bulk lattices)

Stepped surfaces (often consisting of mono-atomic height steps and terraces of arbitrary width) with general Miller indices (hkl) are viewed as composed of alternating low-Miller-index facets $(h_s k_s l_s)$ for the step faces and $(h_t k_t l_t)$ for the terrace planes. The step notation designates such a surface as

$$(hkl) = n(h_t k_t l_t) \times (h_s k_s l_s)$$

where n measures the terrace width in units of atom rows.

> *Example*   The (755) surface of an fcc solid, with (111) oriented terraces and (100) oriented step faces is designated in the step notation as $6(111)\times(100)$.

The microfacet notation allows more complex surface terminations, using decomposition into three independent low-index facet orientations $(h_1 k_1 l_1)$, $(h_2 k_2 l_2)$ and $(h_3 k_3 l_3)$:

$$(hkl) = a_\lambda^1(h_1 k_1 l_1) + a_\mu^2(h_2 k_2 l_2) + a_\nu^3(h_3 k_3 l_3).$$

The factors a^β are simple vectorial decomposition coefficients of (hkl) onto the three microfacet Miller-index vectors, while the subscripts λ, μ and ν indicate the relative facet sizes in terms of 2D unit cells on each facet.

> *Example*   The (10,8,7) surface of an fcc solid, with (111) oriented terraces and steps zigzagging with alternating orientations $(11\bar{1})$ and (100), is designated in the microfacet notation as $(10,8,7) = \left[(15/2)_{15}(111) + (1/2)_1(11\bar{1}) + 2_2(100)\right]$.

(iii) Notations for superlattices at surfaces

Matrix notation: a 2D superlattice with basis vectors $\boldsymbol{b}_1, \boldsymbol{b}_2$ on a bulk-like substrate with 2D surface basis vectors $\boldsymbol{a}_1, \boldsymbol{a}_2$ satisfies the relation $\boldsymbol{b}_1 = m_{11}\boldsymbol{a}_1 + m_{12}\boldsymbol{a}_2$ and $\boldsymbol{b}_2 = m_{21}\boldsymbol{a}_1 + m_{22}\boldsymbol{a}_2$, thus defining a matrix $M = \begin{pmatrix} m_{11} & m_{12} \\ m_{21} & m_{22} \end{pmatrix}$ that uniquely specifies the superlattice. If the matrix elements are all integers, the superlattice is termed commensurate with the substrate lattice.

Wood notation: in many cases the superlattice can be labeled as $i \left(\dfrac{b_1}{a_1} \times \dfrac{b_2}{a_2} \right) R\alpha^\circ$, where i is c for centered superlattices or i is p for primitive superlattices (p is often dropped) and α is a rotation angle relative to the substrate basis vectors (this part is dropped when $\alpha = 0$).

> *Example*   The black and white chess board is a superlattice relative to the underlying set of all squares. This superlattice can be designated with a matrix $\begin{pmatrix} 1 & 1 \\ -1 & 1 \end{pmatrix}$, which is equivalent to the Wood label $(\sqrt{2} \times \sqrt{2})\, R45^\circ$ or with the alternative description c(2×2).

(iv) Symbols

a_1, a_2	2D substrate basis vector
$a_1{}^*, a_2{}^*$	2D reciprocal substrate basis vector
b_1, b_2	2D superlattice basis vectors
$b_1{}^*, b_2{}^*$	2D reciprocal superlattice basis vectors
g	2D reciprocal lattice vector
hk	2D Miller indices
hkl	3D Miller indices
M	matrix for superlattice notation
z	coordinate perpendicular to surface
θ	surface coverage
Θ_D	Debye temperature
$\Delta\phi_W$	work function change
λ_e	electron mean free path
ϕ_W	work function

2.15 TRANSPORT PROPERTIES

The names and symbols recommended here are in agreement with those recommended by IUPAP [4] and ISO [5.m]. Further information on transport phenomena in electrochemical systems can also be found in [104].

The following symbols are used in the definitions: mass (m), time (t), volume (V), area (A), density (ρ), speed (v), length (l), viscosity (η), pressure (p), acceleration of free fall (g), cubic expansion coefficient (α), thermodynamic temperature (T), surface tension (γ), speed of sound (c), mean free path (λ), frequency (f), thermal diffusivity (a), coefficient of heat transfer (h), thermal conductivity (k), specific heat capacity at constant pressure (c_p), diffusion coefficient (D), mole fraction (x), mass transfer coefficient (k_d), permeability (μ), electric conductivity (κ), and magnetic flux density (B).

Name	Symbol	Definition	SI unit	Notes
flux of mass m	q_m	$q_m = dm/dt$	kg s^{-1}	1
flux density of mass m	J_m	$J_m = q_m/A$	kg m^{-2} s^{-1}	2, 3
heat flux, thermal power	Φ, P	$\Phi = dQ/dt$	W	1
heat flux density	J_q	$J_q = \Phi/A$	W m^{-2}	3
thermal conductance	G	$G = \Phi/\Delta T$	W K^{-1}	
thermal resistance	R	$R = 1/G$	K W^{-1}	
thermal conductivity	λ, k	$\lambda = -J_q/(dT/dl)$	W m^{-1} K^{-1}	
coefficient of heat transfer	$h, (k, K, \alpha)$	$h = J_q/\Delta T$	W m^{-2} K^{-1}	
thermal diffusivity	a	$a = \lambda/\rho c_p$	m^2 s^{-1}	
diffusion coefficient	D	$D = -J_n/(dc/dl)$	m^2 s^{-1}	4
viscosity, shear viscosity	η	$\eta = -\tau_{xz}(\partial V_x/\partial z)^{-1}$	Pa s	5
bulk viscosity	κ	$\kappa = -\tau_{xx}(\boldsymbol{\nabla} \cdot \boldsymbol{V})^{-1}$	Pa s	5
thermal diffusion coefficient	D^{T}	$D^{\mathrm{T}} = J_x c^{-1}(dT/dx)^{-1}$	m^2 K^{-1} s^{-1}	

(1) A vector quantity, $\boldsymbol{q}_m = (dm/dt)\,\boldsymbol{e}$, where $\boldsymbol{e}$ is the unit vector in the direction of the flow, can be defined, sometimes called the "flow rate" of m. Similar definitions apply to any extensive quantity.
(2) The flux density of molecules, J_N, determines either the rate at which it would be covered if each molecule stuck, or the rate of effusion through a hole in the surface. In studying the exposure, $\int J_N dt$, of a surface to a gas, surface scientists find it useful to use the product of pressure and time as a measure of the exposure, since this product is equal to the number flux density, J_N, times the time $J_N t = C(\bar{u}/4)pt$, where C is the number density of molecules, $\bar{u}$ their average speed, k the Boltzmann constant and T the thermodynamic temperature. The unit langmuir (symbol L) corresponds to the exposure of a surface to a gas at 10^{-6} Torr for 1 s.
(3) In previous editions, the term "flux" was used for what is now called "flux density", in agreement with IUPAP [4]. The term "density" in the name of an intensive physical quantity usually implies "extensive quantity divided by volume" for scalar quantities but "extensive quantity divided by area" for vector quantities denoting flow or flux.
(4) c is the amount concentration.
(5) See also Section 2.2, p. 15; τ is the shear stress tensor.

2.15.1 Transport characteristic numbers: Quantities of dimension one

Name	Symbol	Definition	Notes
Reynolds number	Re	$Re = \rho v l / \eta$	
Euler number	Eu	$Eu = \Delta p / \rho v^2$	
Froude number	Fr	$Fr = v/(lg)^{1/2}$	
Grashof number	Gr	$Gr = l^3 g \alpha \Delta T \rho^2 / \eta^2$	
Weber number	We	$We = \rho v^2 l / \gamma$	
Mach number	Ma	$Ma = v/c$	
Knudsen number	Kn	$Kn = \lambda / l$	
Strouhal number	Sr	$Sr = lf/v$	
Fourier number	Fo	$Fo = at/l^2$	
Péclet number	Pe	$Pe = vl/a$	
Rayleigh number	Ra	$Ra = l^3 g \alpha \Delta T \rho / \eta a$	
Nusselt number	Nu	$Nu = hl/k$	
Stanton number	St	$St = h/\rho v c_p$	
Fourier number for mass transfer	Fo^*	$Fo^* = Dt/l^2$	1
Péclet number for mass transfer	Pe^*	$Pe^* = vl/D$	1
Grashof number for mass transfer	Gr^*	$Gr^* = l^3 g \left(\dfrac{\partial \rho}{\partial x} \right)_{T,p} \left(\dfrac{\Delta x \rho}{\eta^2} \right)$	1
Nusselt number for mass transfer	Nu^*	$Nu^* = k_\mathrm{d} l / D$	1, 2
Stanton number for mass transfer	St^*	$St^* = k_\mathrm{d}/v$	1
Prandtl number	Pr	$Pr = \eta / \rho a$	
Schmidt number	Sc	$Sc = \eta / \rho D$	
Lewis number	Le	$Le = a/D$	
magnetic Reynolds number	$Rm,\ Re_\mathrm{m}$	$Rm = v \mu \kappa l$	
Alfvén number	Al	$Al = v(\rho \mu)^{1/2}/B$	
Hartmann number	Ha	$Ha = Bl(\kappa/\eta)^{1/2}$	
Cowling number	Co	$Co = B^2 / \mu \rho v^2$	

(1) This quantity applies to the transport of matter in binary mixtures [72].
(2) The name Sherwood number and symbol Sh have been widely used for this quantity.

3 DEFINITIONS AND SYMBOLS FOR UNITS

3.1   THE INTERNATIONAL SYSTEM OF UNITS (SI)

The International System of Units (SI) was adopted by the 11th General Conference of Weights and Measures (CGPM) in 1960 [3]. It is a coherent system of units built from seven *SI base units*, one for each of the seven dimensionally independent base quantities (see Section 1.2, p. 4); they are: metre, kilogram, second, ampere, kelvin, mole, and candela for the base quantities length, mass, time, electric current, thermodynamic temperature, amount of substance, and luminous intensity, respectively. The definitions of the SI base units are given in Section 3.3, p. 87. The *SI derived units* are expressed as products of powers of the base units, analogous to the corresponding relations between physical quantities. If no numerical factors other than 1 are used in the defining equations for derived units, then the derived units defined in this way are called *coherent derived units*. The SI base units and the derived units without any multiple or sub-multiple prefixes form a coherent set of units, and are called the *coherent SI units*.

In the International System of Units there is only one coherent SI unit for each physical quantity. This is either the appropriate SI base unit itself (see Section 3.2, p. 86) or the appropriate SI derived unit (see sections 3.4 and 3.5, p. 89 and 90). However, any of the approved decimal prefixes, called *SI prefixes*, may be used to construct decimal multiples or submultiples of SI units (see Section 3.6, p. 91). The SI units and the decimal multiples and submultiples constructed with the SI prefixes are called the complete set of SI units, or simply the SI units, or the units of the SI.

It is recommended that only units of the SI be used in science and technology (with SI prefixes where appropriate). Where there are special reasons for making an exception to this rule, it is recommended always to define the units in terms of SI units.

3.2   NAMES AND SYMBOLS FOR THE SI BASE UNITS

The symbols listed here are internationally agreed and shall not be changed in other languages or scripts. See sections 1.3 and 1.4, p. 5 and p. 6 on the printing of symbols for units.

Base quantity	SI base unit	
	Name	*Symbol*
length	metre	m
mass	kilogram	kg
time	second	s
electric current	ampere	A
thermodynamic temperature	kelvin	K
amount of substance	mole	mol
luminous intensity	candela	cd

3.3   DEFINITIONS OF THE SI BASE UNITS

The following definitions of the seven SI base units are adopted by the General Conference on Weights and Measures (CGPM) [3].

metre (symbol: m)

> **The metre is the length of path traveled by light in vacuum during a time interval of $1/299\,792\,458$ of a second.**
> (17th CGPM, 1983)

kilogram (symbol: kg)

> **The kilogram is the unit of mass; it is equal to the mass of the international prototype of the kilogram.**
> (3rd CGPM, 1901) [1]

second (symbol: s)

> **The second is the duration of $9\,192\,631\,770$ periods of the radiation corresponding to the transition between the two hyperfine levels of the ground state of the caesium 133 atom.**
> (13th CGPM, 1967)
>
> This definition refers to a caesium atom at rest at a temperature of 0 K.
> (CIPM, 1997)

In this definition it is understood that the Cs atom at a temperature of $T = 0$ K is unperturbed by black-body radiation. The frequency of primary frequency standards should therefore be corrected for the frequency shift due to the ambient radiation, as stated by the Consultative Committee for Time and Frequency (CCTF, 1999).

ampere (symbol: A)

> **The ampere is that constant current which, if maintained in two straight parallel conductors of infinite length, of negligible circular cross-section, and placed 1 metre apart in vacuum, would produce between these conductors a force equal to 2×10^{-7} newton per metre of length.**
> (9th CGPM, 1948)

kelvin (symbol: K)

> **The kelvin, unit of thermodynamic temperature, is the fraction $1/273.16$ of the thermodynamic temperature of the triple point of water.**
> (13th CGPM, 1967)
>
> This definition refers to water having the isotopic composition defined exactly by the following amount-of-substance ratios: $0.000\,155\,76$ mole of ^{2}H per mole of ^{1}H, $0.000\,379\,9$ mole of ^{17}O per mole of ^{16}O, and $0.002\,005\,2$ mole of ^{18}O per mole of ^{16}O.
> (CIPM, 2005) [2]

---

[1] The kilogram is the only base unit which is not defined by a measurement on a system defined by natural microscopic constants or an experimental setup derived from such a system. Rather it is defined by a human artefact (the international prototype of the kilogram). Therefore, alternative definitions of the kilogram are under current discussion [127–129].

[2] See also Section 2.11, note 2, p. 56.

mole (symbol: mol)

> **1. The mole is the amount of substance of a system which contains as many elementary entities as there are atoms in 0.012 kilogram of carbon 12; its symbol is "mol".**
> **2. When the mole is used, the elementary entities must be specified and may be atoms, molecules, ions, electrons, other particles, or specified groups of such particles.**
> (14th CGPM, 1971)

> In this definition, it is understood that unbound atoms of carbon 12, at rest and in their ground state, are referred to.
> (CIPM, 1980)

Examples of the use of the mole

> 1 mol of $^1H_2$ contains about 6.022×10^{23} $^1H_2$ molecules, or 12.044×10^{23} ^1H atoms
> 1 mol of HgCl has a mass of 236.04 g
> 1 mol of Hg_2Cl_2 has a mass of 472.09 g
> 1 mol of Hg_2^{2+} has a mass of 401.18 g and a charge of 192.97 kC
> 1 mol of $Fe_{0.91}S$ has a mass of 82.88 g
> 1 mol of e^- has a mass of 548.58 μg and a charge of -96.49 kC
> 1 mol of photons whose frequency is 5×10^{14} Hz
> has an energy of about 199.5 kJ

Specification of the entity does not imply that the entities are identical: one may have 1 mol of an isotope mixture or gas mixture.

candela (symbol: cd)

> **The candela is the luminous intensity, in a given direction, of a source that emits monochromatic radiation of frequency 540×10^{12} hertz and that has a radiant intensity in that direction of 1/683 watt per steradian.**
> (16th CGPM, 1979)

3.4   SI DERIVED UNITS WITH SPECIAL NAMES AND SYMBOLS

Derived quantity	SI derived unit Name	Symbol	Expressed in terms of other SI units		Notes
plane angle	radian	rad	$m\,m^{-1}$	$=\;1$	1
solid angle	steradian	sr	$m^2\,m^{-2}$	$=\;1$	1
frequency	hertz	Hz	s^{-1}		2
force	newton	N	$m\,kg\,s^{-2}$		
pressure, stress	pascal	Pa	$N\,m^{-2}$	$=\;m^{-1}\,kg\,s^{-2}$	
energy, work, heat	joule	J	$N\,m$	$=\;m^2\,kg\,s^{-2}$	
power, radiant flux	watt	W	$J\,s^{-1}$	$=\;m^2\,kg\,s^{-3}$	
electric charge	coulomb	C	$A\,s$		
electric potential, electromotive force, electric tension	volt	V	$J\,C^{-1}$	$=\;m^2\,kg\,s^{-3}\,A^{-1}$	
electric resistance	ohm	Ω	$V\,A^{-1}$	$=\;m^2\,kg\,s^{-3}\,A^{-2}$	
electric conductance	siemens	S	Ω^{-1}	$=\;m^{-2}\,kg^{-1}\,s^3\,A^2$	
electric capacitance	farad	F	$C\,V^{-1}$	$=\;m^{-2}\,kg^{-1}\,s^4\,A^2$	
magnetic flux	weber	Wb	$V\,s$	$=\;m^2\,kg\,s^{-2}\,A^{-1}$	
magnetic flux density	tesla	T	$Wb\,m^{-2}$	$=\;kg\,s^{-2}\,A^{-1}$	
inductance	henry	H	$V\,A^{-1}\,s$	$=\;m^2\,kg\,s^{-2}\,A^{-2}$	
Celsius temperature	degree Celsius	°C			3
luminous flux	lumen	lm	$cd\,sr$	$=\;cd$	
illuminance	lux	lx	$lm\,m^{-2}$	$=\;cd\,m^{-2}$	
activity, (radioactivity) referred to a radio-nuclide	becquerel	Bq	s^{-1}		4
absorbed dose, kerma	gray	Gy	$J\,kg^{-1}$	$=\;m^2\,s^{-2}$	4
dose equivalent (dose equivalent index)	sievert	Sv	$J\,kg^{-1}$	$=\;m^2\,s^{-2}$	4
catalytic activity	katal	kat	$mol\,s^{-1}$		4, 5

(1) Radian and steradian are derived units. Since they are then of dimension 1, this leaves open the possibility of including them or omitting them in expressions of SI derived units. In practice this means that rad and sr may be used when appropriate and may be omitted if clarity is not lost.

(2) For angular frequency and for angular velocity the unit rad s^{-1}, or simply s^{-1}, should be used, and this may *not* be replaced with Hz. The unit Hz shall be used *only* for frequency in the sense of cycles per second.

(3) The Celsius temperature t is defined by the equation

$$t/°C = T/K - 273.15$$

The SI unit of Celsius temperature is the degree Celsius, °C, which is equal to the kelvin, K. °C shall be treated as a single symbol, with no space between the ° sign and the C. The symbol °K, and the symbol °, shall no longer be used for the unit of thermodynamic temperature.

(4) Becquerel is the basic unit to be used in nuclear and radiochemistry; becquerel, gray, and sievert are admitted for reasons of safeguarding human health [3].

(5) When the amount of a catalyst cannot be expressed as a number of elementary entities, an amount of substance, or a mass, a "catalytic activity" can still be defined as a property of the catalyst measured by a catalyzed rate of conversion under specified optimized conditions. The katal, 1 kat = 1 mol s^{-1}, should replace the "(enzyme) unit", 1 U = µmol $min^{-1} \approx 16.67$ nkat [130].

3.5 SI DERIVED UNITS FOR OTHER QUANTITIES

This table gives examples of other SI derived units; the list is merely illustrative.

Derived quantity	Symbol	SI derived unit Expressed in terms of SI base units	Notes
efficiency	W W^{-1}	$=$ 1	
area	m^2		
volume	m^3		
speed, velocity	m s^{-1}		
angular velocity	rad s^{-1}	$=$ s^{-1}	
acceleration	m s^{-2}		
moment of force	N m	$=$ m^2 kg s^{-2}	
repetency, wavenumber	m^{-1}		1
density, mass density	kg m^{-3}		
specific volume	m^3 kg^{-1}		
amount concentration	mol m^{-3}		2
molar volume	m^3 mol^{-1}		
heat capacity, entropy	J K^{-1}	$=$ m^2 kg s^{-2} K^{-1}	
molar heat capacity, molar entropy	J K^{-1} mol^{-1}	$=$ m^2 kg s^{-2} K^{-1} mol^{-1}	
specific heat capacity, specific entropy	J K^{-1} kg^{-1}	$=$ m^2 s^{-2} K^{-1}	
molar energy	J mol^{-1}	$=$ m^2 kg s^{-2} mol^{-1}	
specific energy	J kg^{-1}	$=$ m^2 s^{-2}	
energy density	J m^{-3}	$=$ m^{-1} kg s^{-2}	
surface tension	N m^{-1}	$=$ kg s^{-2}	
heat flux density, irradiance	W m^{-2}	$=$ kg s^{-3}	
thermal conductivity	W m^{-1} K^{-1}	$=$ m kg s^{-3} K^{-1}	
kinematic viscosity, diffusion coefficient	m^2 s^{-1}		
dynamic viscosity, shear viscosity	Pa s	$=$ m^{-1} kg s^{-1}	
electric charge density	C m^{-3}	$=$ m^{-3} s A	
electric current density	A m^{-2}		
conductivity	S m^{-1}	$=$ m^{-3} kg^{-1} s^3 A^2	
molar conductivity	S m^2 mol^{-1}	$=$ kg^{-1} s^3 A^2 mol^{-1}	
permittivity	F m^{-1}	$=$ m^{-3} kg^{-1} s^4 A^2	
permeability	H m^{-1}	$=$ m kg s^{-2} A^{-2}	
electric field strength	V m^{-1}	$=$ m kg s^{-3} A^{-1}	
magnetic field strength	A m^{-1}		
exposure (X and γ rays)	C kg^{-1}	$=$ kg^{-1} s A	
absorbed dose rate	Gy s^{-1}	$=$ m^2 s^{-3}	

(1) The word "wavenumber" denotes the quantity "reciprocal wavelength". Its widespread use to denote the unit cm^{-1} should be discouraged.

(2) The words "amount concentration" are an abbreviation for "amount-of-substance concentration". When there is not likely to be any ambiguity this quantity may be called simply "concentration".

3.6 SI PREFIXES AND PREFIXES FOR BINARY MULTIPLES

The following prefixes [3] are used to denote decimal multiples and submultiples of SI units.

| Submultiple | *Prefix* | | | *Multiple* | *Prefix* | |
	Name	*Symbol*			*Name*	*Symbol*
10^{-1}	deci	d		10^{1}	deca	da
10^{-2}	centi	c		10^{2}	hecto	h
10^{-3}	milli	m		10^{3}	kilo	k
10^{-6}	micro	μ		10^{6}	mega	M
10^{-9}	nano	n		10^{9}	giga	G
10^{-12}	pico	p		10^{12}	tera	T
10^{-15}	femto	f		10^{15}	peta	P
10^{-18}	atto	a		10^{18}	exa	E
10^{-21}	zepto	z		10^{21}	zetta	Z
10^{-24}	yocto	y		10^{24}	yotta	Y

Prefix symbols shall be printed in Roman (upright) type with no space between the prefix and the unit symbol.

Example   kilometre, km

When a prefix is used with a unit symbol, the combination is taken as a new symbol that can be raised to any power without the use of parentheses.

Examples $1 \text{ cm}^3 = (10^{-2} \text{ m})^3 = 10^{-6} \text{ m}^3$
$1 \text{ μs}^{-1} = (10^{-6} \text{ s})^{-1} = 10^{6} \text{ s}^{-1}$
$1 \text{ V/cm} = 1 \text{ V}/(10^{-2} \text{ m}) = 10^{2} \text{ V/m}$
$1 \text{ mmol/dm}^3 = 10^{-3} \text{ mol}/(10^{-3} \text{ m}^3) = 1 \text{ mol m}^{-3}$

A prefix shall never be used on its own, and prefixes are not to be combined into compound prefixes.

Example   pm, not μμm

The names and symbols of decimal multiples and submultiples of the SI base unit of mass, the kilogram, symbol kg, which already contains a prefix, are constructed by adding the appropriate prefix to the name gram and symbol g.

Examples   mg, not μkg; Mg, not kkg

The International Electrotechnical Commission (IEC) has standardized the following prefixes for binary multiples, mainly used in information technology, to be distinguished from the SI prefixes for decimal multiples [7].

| Multiple | *Prefix* | | Origin |
	Name	*Symbol*	
$(2^{10})^1 = (1024)^1$	kibi	Ki	kilobinary
$(2^{10})^2 = (1024)^2$	mebi	Mi	megabinary
$(2^{10})^3 = (1024)^3$	gibi	Gi	gigabinary
$(2^{10})^4 = (1024)^4$	tebi	Ti	terabinary
$(2^{10})^5 = (1024)^5$	pebi	Pi	petabinary
$(2^{10})^6 = (1024)^6$	exbi	Ei	exabinary
$(2^{10})^7 = (1024)^7$	zebi	Zi	zettabinary
$(2^{10})^8 = (1024)^8$	yobi	Yi	yottabinary

3.7 NON-SI UNITS ACCEPTED FOR USE WITH THE SI

The following units are not part of the SI, but it is recognized by the CGPM [3] that they will continue to be used in appropriate contexts. SI prefixes may be attached to some of these units, such as millilitre, mL; megaelectronvolt, MeV; kilotonne, kt. A more extensive list of non-SI units, with conversion factors to the corresponding SI units, is given in Chapter 7, p. 129.

| Physical quantity | Unit accepted for use with the SI | | | Notes |
	Name	Symbol	Value in SI units	
time	minute	min	$= 60$ s	
time	hour	h	$= 3600$ s	
time	day	d	$= 86\ 400$ s	
plane angle	degree	°, deg	$= (\pi/180)$ rad	
plane angle	minute	$'$	$= (\pi/10\ 800)$ rad	
plane angle	second	$''$	$= (\pi/648\ 000)$ rad	
volume	litre	l, L	$= 1\ \mathrm{dm}^3 = 10^{-3}\ \mathrm{m}^3$	1
mass	tonne	t	$= 1\ \mathrm{Mg} = 10^3$ kg	
level of a field quantity, level of a power quantity	neper	Np	$= \ln \mathrm{e} = (1/2)\ln \mathrm{e}^2 = 1$	2
level of a field quantity, level of a power quantity	bel	B		2
energy	electronvolt	eV $(= e \cdot 1$ V$)$	$= 1.602\ 176\ 487(40) \times 10^{-19}$ J	3
mass	dalton, unified atomic mass unit	Da, u $(= m_\mathrm{a}(^{12}\mathrm{C})/12)$	$= 1.660\ 538\ 782(83) \times 10^{-27}$ kg, $= 1$ Da	3, 4
length	nautical mile	M	$= 1852$ m	5
	astronomical unit	ua	$= 1.495\ 978\ 706\ 91(6) \times 10^{11}$ m	6

(1) The alternative symbol L is the only exception of the general rule that symbols for units shall be printed in lower case letters unless they are derived from a personal name. In order to avoid the risk of confusion between the letter l and the number 1, the use of L is accepted. However, only the lower case l is used by ISO and IEC.

(2) For logarithmic ratio quantities and their units, see [131].

(3) The values of these units in terms of the corresponding SI units are not exact, since they depend on the values of the physical constants e (for electronvolt) and $m_\mathrm{a}(^{12}\mathrm{C})$ or N_A (for the unified atomic mass unit), which are determined by experiment, see Chapter 5, p. 111.

(4) The dalton, with symbol Da, and the unified atomic mass unit, with symbol u, are alternative names for the same unit. The dalton may be combined with SI prefixes to express the masses of large or small entities.

(5) There is no agreed symbol for the nautical mile. The SI Brochure uses the symbol M.

(6) The astronomical unit is a unit of length approximately equal to the mean Earth-Sun distance. Its value is such that, when used to describe the motion of bodies in the Solar System, the heliocentric gravitational constant is $(0.017\ 202\ 098\ 95)^2\ \mathrm{ua}^3\ \mathrm{d}^{-2}$ (see also [3]).

3.8   COHERENT UNITS AND CHECKING DIMENSIONS

If equations between numerical values have the same form as equations between physical quantities, then the system of units defined in terms of base units avoids numerical factors between units, and is said to be a *coherent system*. For example, the kinetic energy T of a particle of mass m moving with a speed v is defined by the equation

$$T = (1/2)\, mv^2$$

but the SI unit of kinetic energy is the joule, defined by the equation

$$J = kg\,(m/s)^2 = kg\,m^2\,s^{-2}$$

where it is to be noted that the factor $(1/2)$ is omitted. In fact the joule, symbol J, is simply a special name and symbol for the product of units $kg\,m^2\,s^{-2}$.

The International System (SI) is a coherent system of units. The advantage of a coherent system of units is that if the value of each quantity is substituted for the quantity symbol in any quantity equation, then the units may be canceled, leaving an equation between numerical values which is exactly similar (including all numerical factors) to the original equation between the quantities. Checking that the units cancel in this way is sometimes described as checking the dimensions of the equation.

The use of a coherent system of units is not essential. In particular the use of multiple or submultiple prefixes destroys the coherence of the SI, but is nonetheless often convenient.

3.9 FUNDAMENTAL PHYSICAL CONSTANTS USED AS UNITS

Sometimes fundamental physical constants, or other well defined physical quantities, are used as though they were units in certain specialized fields of science. For example, in astronomy it may be more convenient to express the mass of a star in terms of the mass of the sun, and to express the period of the planets in their orbits in terms of the period of the earth's orbit, rather than to use SI units. In atomic and molecular physics it is similarly more convenient to express masses in terms of the electron mass, m_e, or in terms of the unified atomic mass unit, $1\ u = m_u = m(^{12}C/12)$, and to express charges in terms of the elementary charge e, rather than to use SI units. One reason for using such physical quantities as though they were units is that the nature of the experimental measurements or calculations in the specialized field may be such that results are naturally obtained in such terms, and can only be converted to SI units at a later stage. When physical quantities are used as units in this way their relation to the SI must be determined by experiment, which is subject to uncertainty, and the conversion factor may change as new and more precise experiments are developed. Another reason for using such units is that uncertainty in the conversion factor to the SI may be greater than the uncertainty in the ratio of the measurements expressed in terms of the physical constant as a unit. Both reasons make it preferable to present experimental results without converting to SI units.

Three such physical quantities that have been recognized as units by the CIPM are the electronvolt (eV), the dalton (Da) or the unified atomic mass unit (u), and the astronomical unit (ua), listed below [3]. The electronvolt is the product of a fundamental constant (the elementary charge, e) and the SI unit of potential difference (the volt, V). The dalton is related to the mass of the carbon-12 nuclide, and is thus a fundamental constant. The astronomical unit is a more arbitrarily defined constant that is convenient to astronomers. However, there are many other physical quantities or fundamental constants that are sometimes used in this way as though they were units, so that it is hardly possible to list them all.

Physical quantity	Name of unit	Symbol for unit	Value in SI units	Notes
energy	electronvolt	eV	$1\ eV = 1.602\ 176\ 487(40) \times 10^{-19}$ J	1
mass	dalton, unified atomic mass unit	Da, u	$1\ Da = 1.660\ 538\ 782(83) \times 10^{-27}$ kg	2
length	astronomical unit	ua	$1\ ua = 1.495\ 978\ 706\ 91(6) \times 10^{11}$ m	3

(1) The electronvolt is the kinetic energy acquired by an electron in passing through a potential barrier of 1 V in vacuum.

(2) The dalton and the unified atomic mass unit are alternative names for the same unit. The dalton may be combined with the SI prefixes to express the masses of large molecules in kilodalton (kDa) or megadalton (MDa).

(3) The value of the astronomical unit in SI units is defined such that, when used to describe the motion of bodies in the solar system, the heliocentric gravitational constant is $(0.017\ 202\ 098\ 95)^2$ $ua^3\ d^{-2}$. The value must be obtained from experiment, and is therefore not known exactly (see also [3]).

3.9.1 Atomic units [22] (see also Section 7.3, p. 143)

One particular group of physical constants that are used as though they were units deserve special mention. These are the so-called *atomic units* and arise in calculations of electronic wavefunctions for atoms and molecules, i.e. in quantum chemistry. The first five atomic units in the table below have special names and symbols. Only four of these are independent; all others may be derived by multiplication and division in the usual way, and the table includes a number of examples. The relation between the five named atomic units may be expressed by any one of the equations

$$E_h = \hbar^2/m_e a_0{}^2 = e^2/4\pi\varepsilon_0 a_0 = m_e e^4/(4\pi\varepsilon_0)^2\hbar^2$$

The relation of atomic units to the corresponding SI units involves the values of the fundamental physical constants, and is therefore not exact. The numerical values in the table are from the CODATA compilation [23] and based on the fundamental constants given in Chapter 5, p. 111. The numerical results of calculations in theoretical chemistry are frequently quoted in atomic units, or as numerical values in the form *physical quantity* divided by *atomic unit*, so that the reader may make the conversion using the current best estimates of the physical constants.

Physical quantity	Name of unit	Symbol for unit	Value in SI units	Notes
mass	electron mass	m_e	$= 9.109\ 382\ 15(45)\times10^{-31}$ kg	
charge	elementary charge	e	$= 1.602\ 176\ 487(40)\times10^{-19}$ C	
action, (angular momentum)	Planck constant divided by 2π	$\hbar$	$= 1.054\ 571\ 628(53)\times10^{-34}$ J s	1
length	Bohr radius	a_0	$= 5.291\ 772\ 085\ 9(36)\times10^{-11}$ m	1
energy	Hartree energy	E_h	$= 4.359\ 743\ 94(22)\times10^{-18}$ J	1
time		$\hbar/E_h$	$= 2.418\ 884\ 326\ 505(16)\times10^{-17}$ s	
speed		$a_0 E_h/\hbar$	$= 2.187\ 691\ 254\ 1(15)\times10^{6}$ m s^{-1}	2
force		E_h/a_0	$= 8.238\ 722\ 06(41)\times10^{-8}$ N	
linear momentum		$\hbar/a_0$	$= 1.992\ 851\ 565(99)\times10^{-24}$ N s	
electric current		$eE_h/\hbar$	$= 6.623\ 617\ 63(17)\times10^{-3}$ A	
electric field		E_h/ea_0	$= 5.142\ 206\ 32(13)\times10^{11}$ V m^{-1}	
electric dipole moment		ea_0	$= 8.478\ 352\ 81(21)\times10^{-30}$ C m	
electric quadrupole moment		$ea_0{}^2$	$= 4.486\ 551\ 07(11)\times10^{-40}$ C m^2	
electric polarizability		$e^2 a_0{}^2/E_h$	$= 1.648\ 777\ 253\ 6(34)\times10^{-41}$ C^2 m^2 J^{-1}	
1st hyper-polarizability		$e^3 a_0{}^3/E_h^2$	$= 3.206\ 361\ 533(81)\times10^{-53}$ C^3 m^3 J^{-2}	
2nd hyper-polarizability		$e^4 a_0{}^4/E_h^3$	$= 6.235\ 380\ 95(31)\times10^{-65}$ C^4 m^4 J^{-3}	
magnetic flux density		$\hbar/ea_0{}^2$	$= 2.350\ 517\ 382(59)\times10^{5}$ T	
magnetic dipole moment		$e\hbar/m_e$	$= 1.854\ 801\ 830(46)\times10^{-23}$ J T^{-1}	3
magnetizability		$e^2 a_0{}^2/m_e$	$= 7.891\ 036\ 433(27)\times10^{-29}$ J T^{-2}	

(1) $\hbar = h/2\pi$; $a_0 = 4\pi\varepsilon_0\hbar^2/m_e e^2$; $E_h = \hbar^2/m_e a_0{}^2$.

(2) The numerical value of the speed of light, when expressed in atomic units, is equal to the reciprocal of the fine-structure constant α;

$c/$(au of speed) $= c\hbar/a_0 E_h = \alpha^{-1} = 137.035\ 999\ 679(94)$.

(3) The atomic unit of magnetic dipole moment is twice the Bohr magneton, μ_B.

3.9.2 The equations of quantum chemistry expressed in terms of reduced quantities using atomic units

It is customary to write the equations of quantum chemistry in terms of reduced quantities. Thus energies are expressed as reduced energies E^*, distances as reduced distances r^*, masses as reduced masses m^*, charges as reduced charges Q^*, and angular momenta as reduced angular momenta J^*, where the reduced quantities are given by the equations

$$E^* = E/E_h, \ r^* = r/a_0, \ m^* = m/m_e, \ Q^* = Q/e, \quad \text{and} \quad J^* = J/\hbar \tag{1}$$

The reduced quantity in each case is the dimensionless ratio of the actual quantity to the corresponding atomic unit. The advantage of expressing all the equations in terms of reduced quantities is that the equations are simplified since all the physical constants disappear from the equations (although this simplification is achieved at the expense of losing the advantage of dimensional checking, since all reduced quantities are dimensionless). For example the Schrödinger equation for the hydrogen atom, expressed in the usual physical quantities, has the form

$$-(\hbar^2/2m_e)\boldsymbol{\nabla}_r^2\,\psi(r,\theta,\phi) + V(r)\psi(r,\theta,\phi) = E\psi(r,\theta,\phi) \tag{2}$$

Here r, θ, and ϕ are the coordinates of the electron, and the operator $\boldsymbol{\nabla}_r$ involves derivatives $\partial/\partial r, \partial/\partial\theta$, and $\partial/\partial\phi$. However in terms of reduced quantities the corresponding equation has the form

$$-(1/2)\boldsymbol{\nabla}_{r^*}^2\,\psi(r^*,\theta,\phi) + V^*(r^*)\psi(r^*,\theta,\phi) = E^*\psi(r^*,\theta,\phi) \tag{3}$$

where $\boldsymbol{\nabla}_{r^*}$ involves derivatives $\partial/\partial r^*, \partial/\partial\theta$, and $\partial/\partial\phi$. This may be shown by substituting the reduced (starred) quantities for the actual (unstarred) quantities in Equation (2) which leads to Equation (3).

In the field of quantum chemistry it is customary to write all equations in terms of reduced (starred) quantities, so that all quantities become dimensionless, and all fundamental constants such as $e, m_e, \hbar, E_h$, and a_0 disappear from the equations. As observed above this simplification is achieved at the expense of losing the possibility of dimensional checking. To compare the results of a numerical calculation with experiment it is of course necessary to transform the calculated values of the reduced quantities back to the values of the actual quantities using Equation (1).

Unfortunately it is also customary not to use the star that has been used here, but instead to use exactly the same symbol for the dimensionless reduced quantities and the actual quantities. This makes it impossible to write equations such as (1). (It is analogous to the situation that would arise if we were to use exactly the same symbol for h and $\hbar$, where $\hbar = h/2\pi$, thus making it impossible to write the relation between h and $\hbar$.) This may perhaps be excused on the grounds that it becomes tedious to include a star on the symbol for every physical quantity when writing the equations in quantum chemistry, but it is important that readers unfamiliar with the field should realize what has been done. It is also important to realize how the values of quantities "expressed in atomic units", i.e. the values of reduced quantities, may be converted back to the values of the original quantities in SI units.

It is also customary to make statements such as "in atomic units $e, m_e, \hbar, E_h$, and a_0 are all equal to 1", which is not a correct statement. The correct statement would be that in atomic units the elementary charge is equal to 1 e, the mass of an electron is equal to 1 m_e, etc. The difference between equations such as (3), which contain no fundamental constants, and (2) which do contain fundamental constants, concerns the quantities rather than the units. In (3) all the quantities are dimensionless reduced quantities, defined by (1), whereas in (2) the quantities are the usual (dimensioned) physical quantities with which we are familiar in other circumstances.

Finally, many authors make no use of the symbols for the atomic units listed in the tables above, but instead use the symbol "a.u." or "au" for all atomic units. This custom should not be followed. It leads to confusion, just as it would if we were to write "SI" as a symbol for every SI unit, or "CGS" as a symbol for every CGS unit.

Examples   For the hydrogen molecule the equilibrium bond length r_e, and the
dissociation energy D_e, are given by
$r_e = 2.1\ a_0$ *not* $r_e = 2.1$ a.u.
$D_e = 0.16\ E_h$ *not* $D_e = 0.16$ a.u.

3.10 DIMENSIONLESS QUANTITIES

Values of dimensionless physical quantities, more properly called "quantities of dimension one", are often expressed in terms of mathematically exactly defined values denoted by special symbols or abbreviations, such as % (percent). These symbols are then treated as units, and are used as such in calculations.

3.10.1 Fractions (relative values, yields, and efficiencies)

Fractions such as relative uncertainty, amount-of-substance fraction x (also called amount fraction), mass fraction w, and volume fraction φ (see Section 2.10, p. 47 for all these quantities), are sometimes expressed in terms of the symbols in the table below.

Name	Symbol	Value	Example
percent	%	10^{-2}	The isotopic abundance of carbon-13 expressed as an amount-of-substance fraction is $x = 1.1$ %.
permille	‰	10^{-3}	The mass fraction of water in a sample is $w = 2.3$ ‰.

These multiples of the unit one are not part of the SI and ISO recommends that these symbols should never be used. They are also frequently used as units of "concentration" without a clear indication of the type of fraction implied, e.g. amount-of-substance fraction, mass fraction or volume fraction. To avoid ambiguity they should be used only in a context where the meaning of the quantity is carefully defined. Even then, the use of an appropriate SI unit ratio may be preferred.

Examples   The mass fraction $w = 1.5 \times 10^{-6} = 1.5$ mg/kg.
The amount-of-substance fraction $x = 3.7 \times 10^{-2} = 3.7$ % or $x = 37$ mmol/mol.
Atomic absorption spectroscopy shows the aqueous solution to contain a mass concentration of nickel $\rho(\text{Ni}) = 2.6$ mg dm^{-3}, which is approximately equivalent to a mass fraction $w(\text{Ni}) = 2.6 \times 10^{-6}$.

Note the importance of using the recommended name and symbol for the quantity in each of the above examples. Statements such as "the concentration of nickel was 2.6×10^{-6}" are ambiguous and should be avoided.

The last example illustrates the approximate equivalence of $\rho/\text{mg dm}^{-3}$ and $w/10^{-6}$ in aqueous solution, which follows from the fact that the mass density of a dilute aqueous solution is always approximately 1.0 g cm^{-3}. Dilute solutions are often measured or calibrated to a known mass concentration in mg dm^{-3}, and this unit is then to be preferred to using ppm (or other corresponding abbreviations, which are language dependent) to specify a mass fraction.

3.10.2 Deprecated usage

Adding extra labels to % and similar symbols, such as % (V/V) (meaning % by volume) should be avoided. Qualifying labels may be added to symbols for physical quantities, but never to units.

Example   A mass fraction $w = 0.5$ %, but *not* 0.5 % (m/m).

The symbol % should not be used in combination with other units. In table headings and in labeling the axes of graphs the use of % in the denominator is to be avoided. Although one would write $x(^{13}\text{C}) = 1.1$ %, the notation $100\,x$ is to be preferred to $x/\%$ in tables and graphs (see for example Section 6.3, column 5, p. 122).

The further symbols listed in the table below are also found in the literature, but their use is not recommended. Note that the names and symbols for 10^{-9} and 10^{-12} in this table are here based on the American system of names. In other parts of the world, a billion often stands for 10^{12} and a trillion for 10^{18}. Note also that the symbol ppt is sometimes used for part per thousand, and sometimes for part per trillion. In 1948 the word billion had been proposed for 10^{12} and trillion for 10^{18} [132]. Although ppm, ppb, ppt and alike are widely used in various applications of analytical and environmental chemistry, it is suggested to abandon completely their use because of the ambiguities involved. These units are unnecessary and can be easily replaced by SI-compatible quantities such as pmol/mol (picomole per mole), which are unambiguous. The last column contains suggested replacements (similar replacements can be formulated as mg/g, µg/g, pg/g etc.).

Name	Symbol	Value	Examples	Replacement
part per hundred	pph, %	10^{-2}	The degree of dissociation is 1.5 %.	
part per thousand, permille [1]	ppt, ‰	10^{-3}	An approximate preindustrial value of the CO_2 content of the Earth's atmosphere was 0.275 ‰ (0.275 ppt).	mmol/mol
			The element Ti has a mass fraction 5.65 ‰ (5.65×10^3 ppm) in the Earth's crust.	mg/g
part per million	ppm	10^{-6}	The volume fraction of helium is 20 ppm.	µmol/mol
part per hundred million	pphm	10^{-8}	The mass fraction of impurity in the metal was less than 5 pphm.	
part per billion	ppb	10^{-9}	The air quality standard for ozone is a volume fraction of $\varphi = 120$ ppb.	nmol/mol
part per trillion	ppt	10^{-12}	The natural background volume fraction of NO in air was found to be $\varphi = 140$ ppt.	pmol/mol
part per quadrillion	ppq	10^{-15}		fmol/mol

[1] The permille is also spelled per mill, permill, per mil, permil, per mille, or promille.

3.10.3 Units for logarithmic quantities: neper, bel, and decibel

In some fields, especially in acoustics and telecommunications, special names are given to the number 1 when expressing physical quantities defined in terms of the logarithm of a ratio [131]. For a damped linear oscillation the amplitude of a quantity as a function of time is given by

$$F(t) = A\, e^{-\delta t} \cos \omega t = A\, \mathrm{Re}[e^{(-\delta + i\omega)t}]$$

From this relation it is clear that the coherent SI unit for the decay coefficient δ and the angular frequency ω is the second to the power of minus one, s^{-1}. However, the special names neper, Np, and radian, rad (see Section 2.1, p. 13, Section 3.4, p. 89, and Section 3.7, p. 92), are used for the units of the dimensionless products δt and ωt, respectively. Thus the quantities δ and ω may be expressed in the units Np/s and rad/s, respectively. Used in this way the neper, Np, and the radian, rad, may both be thought of as special names for the number 1.

In the fields of acoustics and signal transmission, signal power levels and signal amplitude levels (or field level) are usually expressed as the decadic or the napierian logarithm of the ratio of the power P to a reference power P_0, or of the field F to a reference field F_0. Since power is often proportional to the square of the field or amplitude (when the field acts on equal impedances in linear systems) it is convenient to define the power level and the field level to be equal in such a case. This is done by defining the field level and the power level according to the relations

$$L_F = \ln(F/F_0), \qquad \text{and} \qquad L_P = (1/2)\ln(P/P_0)$$

so that if $(P/P_0) = (F/F_0)^2$ then $L_P = L_F$. The above equations may be written in the form

$$L_F = \ln(F/F_0) \text{ Np}, \quad \text{and} \quad L_P = (1/2)\ln(P/P_0) \text{ Np}$$

The bel, B, and its more frequently used submultiple the decibel, dB, are used when the field and power levels are calculated using decadic logarithms according to the relations

$$L_P = \lg(P/P_0) \text{ B} = 10 \lg(P/P_0) \text{ dB}$$

and

$$L_F = 2 \lg(F/F_0) \text{ B} = 20 \lg(F/F_0) \text{ dB}$$

The relation between the bel and the neper follows from comparing these equations with the preceeding equations. We obtain

$$L_F = \ln(F/F_0) \text{ Np} = 2 \lg(F/F_0) \text{ B} = \ln(10) \lg(F/F_0) \text{ Np}$$

giving

$$1 \text{ B} = 10 \text{ dB} = (1/2) \ln(10) \text{ Np} \approx 1.151\ 293 \text{ Np}$$

In practice the bel is hardly ever used. Only the decibel is used, to represent the decadic logarithm, particularly in the context of acoustics, and in labeling the controls of power amplifiers. Thus the statement $L_P = n$ dB implies that $10 \lg(P/P_0) = n$.

The general use of special units for logarithmic quantities is discussed in [131]. The quantities power level and field level, and the units bel, decibel and neper, are given in the table and notes that follow.

Name	Quantity	Numerical value multiplied by unit	Notes
field level	$L_F = \ln(F/F_0)$	$= \ln(F/F_0) \text{ Np} = 2 \lg(F/F_0) \text{ B} = 20 \lg(F/F_0) \text{ dB}$	1–3
power level	$L_P = (1/2)\ln(P/P_0)$	$= (1/2)\ln(P/P_0) \text{ Np} = \lg(P/P_0) \text{ B} = 10 \lg(P/P_0) \text{ dB}$	4–6

(1) F_0 is a reference field quantity, which should be specified.
(2) In the context of acoustics the field level is called the sound pressure level and given the symbol L_p, and the reference pressure $p_0 = 20$ μPa.
(3) For example, when $L_F = 1$ Np, $F/F_0 = \mathrm{e} \approx 2.718\ 281\ 8$.
(4) P_0 is a reference power, which should be specified. The factor $1/2$ is included in the definition to make $L_P \widehat{=} L_F$.
(5) In the context of acoustics the power level is called the sound power level and given the symbol L_W, and the reference power $P_0 = 1$ pW.
(6) For example, when $L_P = 1$ B $= 10$ dB, $P/P_0 = 10$; and when $L_P = 2$ B $= 20$ dB, $P/P_0 = 100$; etc.

99

4 RECOMMENDED MATHEMATICAL SYMBOLS

4.1 PRINTING OF NUMBERS AND MATHEMATICAL SYMBOLS [5.a]

1. Numbers in general shall be printed in Roman (upright) type. The decimal sign between digits in a number should be a point (e.g. 2.3) or a comma (e.g. 2,3). When the decimal sign is placed before the first significant digit of a number, a zero shall always precede the decimal sign. To facilitate the reading of long numbers the digits may be separated into groups of three about the decimal sign, using only a thin space (but never a point or a comma, nor any other symbol). However, when there are only four digits before or after the decimal marker we recommend that no space is required and no space should be used.

   Examples   2573.421 736 or 2573,421 736 or $0.257\ 342\ 173\ 6 \times 10^4$ or
                  $0,257\ 342\ 173\ 6 \times 10^4$
                  32 573.4215   or   32 573,4217

2. Numerical values of physical quantities which have been experimentally determined are usually subject to some uncertainty. The experimental uncertainty should always be specified. The magnitude of the uncertainty may be represented as follows

   Examples $l = [5.3478 - 0.0064, 5.3478 + 0.0064]$ cm
                  $l = 5.3478(32)$ cm

   In the first example the range of uncertainty is indicated directly as $[a - b, a + b]$. It is recommended that this notation should be used only with the meaning that the interval $[a - b, a + b]$ contains the true value with a high degree of certainty, such that $b \geqslant 2\sigma$, where σ denotes the standard uncertainty or standard deviation (see Chapter 8, p. 149).

   In the second example, $a(c)$, the range of uncertainty c indicated in parentheses is assumed to apply to the least significant digits of a. It is recommended that this notation be reserved for the meaning that b represents 1σ in the final digits of a.

3. Letter symbols for mathematical constants (e.g. e, π, i $= \sqrt{-1}$) shall be printed in Roman (upright) type, but letter symbols for numbers other than constants (e.g. quantum numbers) should be printed in italic (sloping) type, similar to physical quantities.

4. Symbols for specific mathematical functions and operators (e.g. lb, ln, lg, exp, sin, cos, d, δ, Δ, ∇, ...) shall be printed in Roman type, but symbols for a general function (e.g. $f(x)$, $F(x, y)$, ...) shall be printed in italic type.

5. The operator p (as in pa_{H^+}, $pK = - \lg K$ etc., see Section 2.13.1 (viii), p. 75) shall be printed in Roman type.

6. Symbols for symmetry species in group theory (e.g. S, P, D, ..., s, p, d, ..., Σ, Π, Δ, ..., A_{1g}, B_2'', ...) shall be printed in Roman (upright) type when they represent the state symbol for an atom or a molecule, although they are often printed in italic type when they represent the symmetry species of a point group.

7. Vectors and matrices shall be printed in bold italic type.

   Examples   force $\boldsymbol{F}$, electric field $\boldsymbol{E}$, position vector $\boldsymbol{r}$

Ordinary italic type is used to denote the magnitude of the corresponding vector.

   Example $r = |\boldsymbol{r}|$

Tensor quantities may be printed in bold face italic sans-serif type.

   Examples $\boldsymbol{\mathsf{S}}$, $\boldsymbol{\mathsf{T}}$

Vectors may alternatively be characterized by an arrow, $\vec{A}$, $\vec{a}$ and second-rank tensors by a double arrow, $\overset{\leftrightarrow}{S}, \overset{\leftrightarrow}{T}$.

104

4.2 SYMBOLS, OPERATORS, AND FUNCTIONS [5.k]

Description	Symbol	Notes
signs and symbols		
equal to	$=$	
not equal to	$\neq$	
identically equal to	$\equiv$	
equal by definition to	$\stackrel{\mathrm{def}}{=}$, $:=$	
approximately equal to	$\approx$	
asymptotically equal to	$\simeq$	
corresponds to	$\hat{=}$	
proportional to	$\sim$, $\propto$	
tends to, approaches	$\rightarrow$	
infinity	∞	
less than	$<$	
greater than	$>$	
less than or equal to	$\leqslant$	
greater than or equal to	$\geqslant$	
much less than	$\ll$	
much greater than	$\gg$	
operations		
plus	$+$	
minus	$-$	
plus or minus	$\pm$	
minus or plus	$\mp$	
a multiplied by b	$a\,b$, ab, $a \cdot b$, $a \times b$	1
a divided by b	a/b, ab^{-1}, $\dfrac{a}{b}$	2
magnitude of a	$|a|$	
a to the power n	a^n	
square root of a, and of $a^2 + b^2$	$\sqrt{a}$, $a^{1/2}$, and $\sqrt{a^2 + b^2}$, $\left(a^2 + b^2\right)^{1/2}$	
nth root of a	$a^{1/n}$, $\sqrt[n]{a}$	
mean value of a	$\langle a \rangle$, $\bar{a}$	
sign of a (equal to $a/|a|$ if $a \neq 0$, 0 if $a = 0$)	$\mathrm{sgn}\ a$	
n factorial	$n!$	
binominal coefficient, $n!/p!(n-p)!$	C_p^n, $\binom{n}{p}$	
sum of a_i	$\sum a_i$, $\sum_i a_i$, $\sum_{i=1}^{n} a_i$	
product of a_i	$\prod a_i$, $\prod_i a_i$, $\prod_{i=1}^{n} a_i$	
functions		
sine of x	$\sin x$	
cosine of x	$\cos x$	
tangent of x	$\tan x$	
cotangent of x	$\cot x$	

(1) When multiplication is indicated by a dot, the dot shall be half high: $a \cdot b$.

(2) $a : b$ is as well used for "divided by". However, this symbol is mainly used to express ratios such as length scales in maps.

Description	Symbol	Notes
arc sine of x	$\arcsin x$	3
arc cosine of x	$\arccos x$	3
arc tangent of x	$\arctan x$	3
arc cotangent of x	$\operatorname{arccot} x$	3
hyperbolic sine of x	$\sinh x$	
hyperbolic cosine of x	$\cosh x$	
hyperbolic tangent of x	$\tanh x$	
hyperbolic cotangent of x	$\coth x$	
area hyperbolic sine of x	$\operatorname{arsinh} x$	3
area hyperbolic cosine of x	$\operatorname{arcosh} x$	3
area hyperbolic tangent of x	$\operatorname{artanh} x$	3
area hyperbolic cotangent of x	$\operatorname{arcoth} x$	3
base of natural logarithms	e	
exponential of x	$\exp x$, e^x	
logarithm to the base a of x	$\log_a x$	4
natural logarithm of x	$\ln x$, $\log_e x$	4
logarithm to the base 10 of x	$\lg x$, $\log_{10} x$	4
logarithm to the base 2 of x	$\operatorname{lb} x$, $\log_2 x$	4
greatest integer $\leqslant x$	$\operatorname{ent} x$	
integer part of x	$\operatorname{int} x$	
integer division	$\operatorname{int}(n/m)$	
remainder after integer division	$n/m - \operatorname{int}(n/m)$	
change in x	$\Delta x = x(\text{final}) - x(\text{initial})$	
infinitesimal variation of f	δf	
limit of $f(x)$ as x tends to a	$\lim\limits_{x \to a} f(x)$	
1st derivative of f	$\mathrm{d}f/\mathrm{d}x$, f', $(\mathrm{d}/\mathrm{d}x)f$	
2nd derivative of f	$\mathrm{d}^2 f/\mathrm{d}x^2$, f''	
nth derivative of f	$\mathrm{d}^n f/\mathrm{d}x^n$, $f^{(n)}$	
partial derivative of f	$\partial f/\partial x$, $\partial_x f$, $\mathrm{D}_x f$	
total differential of f	$\mathrm{d}f$	
inexact differential of f	$\mathrm{d}f$	5
first derivative of x with respect to time	$\dot{x}$, $\mathrm{d}x/\mathrm{d}t$	
integral of $f(x)$	$\int f(x)\,\mathrm{d}x$, $\int \mathrm{d}x\, f(x)$	
Kronecker delta	$\delta_{ij} = \begin{cases} 1 & \text{if } i = j \\ 0 & \text{if } i \neq j \end{cases}$	
Levi-Civita symbol	$\varepsilon_{ijk} = \begin{cases} 1 & \text{if } ijk \text{ is a cyclic permutation of 123} \\ & \varepsilon_{123} = \varepsilon_{231} = \varepsilon_{312} = 1 \\ -1 & \text{if } ijk \text{ is an anticyclic permutation of 123} \\ & \varepsilon_{132} = \varepsilon_{321} = \varepsilon_{213} = -1 \\ 0 & \text{otherwise} \end{cases}$	
Dirac delta function (distribution)	$\delta(x)$, $\int f(x)\delta(x)\,\mathrm{d}x = f(0)$	

(3) These are the inverse of the parent function, i.e. $\arcsin x$ is the operator inverse of $\sin x$.
(4) For positive x.
(5) Notation used in thermodynamics, see Section 2.11, note 1, p. 56.

Description	Symbol	Notes
unit step function, Heaviside function	$\varepsilon(x)$, $H(x)$, $h(x)$, $\varepsilon(x) = 1$ for $x > 0$, $\quad \varepsilon(x) = 0$ for $x < 0$.	
gamma function	$\Gamma(x) = \int\limits_0^\infty t^{x-1}e^{-t}\mathrm{d}t$ $\Gamma(n+1) = (n)!$ for positive integers n	
convolution of functions f and g	$f * g = \int\limits_{-\infty}^{+\infty} f(x - x')g(x')\,\mathrm{d}x'$	

complex numbers

Description	Symbol	Notes
square root of -1, $\sqrt{-1}$	i	
real part of $z = a + \mathrm{i}\,b$	$\mathrm{Re}\ z = a$	
imaginary part of $z = a + \mathrm{i}\,b$	$\mathrm{Im}\ z = b$	
modulus of $z = a + \mathrm{i}\,b$, absolute value of $z = a + \mathrm{i}\,b$	$|z| = \left(a^2 + b^2\right)^{1/2}$	
argument of $z = a + \mathrm{i}\,b$	$\arg z$; $\tan(\arg z) = b/a$	
complex conjugate of $z = a + \mathrm{i}\,b$	$z^* = a - \mathrm{i}\,b$	

vectors

Description	Symbol	Notes
vector $\boldsymbol{a}$	$\boldsymbol{a}$, $\vec{a}$	
cartesian components of $\boldsymbol{a}$	a_x, a_y, a_z	
unit vectors in cartesian coordinate system	$\boldsymbol{e}_x$, $\boldsymbol{e}_y$, $\boldsymbol{e}_z$ or $\boldsymbol{i}$, $\boldsymbol{j}$, $\boldsymbol{k}$	
scalar product	$\boldsymbol{a} \cdot \boldsymbol{b}$	
vector or cross product	$\boldsymbol{a} \times \boldsymbol{b}$, $(\boldsymbol{a} \wedge \boldsymbol{b})$	
nabla operator, del operator	$\boldsymbol{\nabla} = \boldsymbol{e}_x \partial/\partial x + \boldsymbol{e}_y \partial/\partial y + \boldsymbol{e}_z \partial/\partial z$	
Laplacian operator	$\boldsymbol{\nabla}^2$, $\triangle = \partial^2/\partial x^2 + \partial^2/\partial y^2 + \partial^2/\partial z^2$	
gradient of a scalar field V	$\boldsymbol{grad}\ V$, $\boldsymbol{\nabla} V$	
divergence of a vector field $\boldsymbol{A}$	$\boldsymbol{div}\ \boldsymbol{A}$, $\boldsymbol{\nabla} \cdot \boldsymbol{A}$	
rotation of a vector field $\boldsymbol{A}$	$\boldsymbol{rot}\ \boldsymbol{A}$, $\boldsymbol{\nabla} \times \boldsymbol{A}$, $(\boldsymbol{curl}\ \boldsymbol{A})$	

matrices

Description	Symbol	Notes
matrix of element A_{ij}	$\boldsymbol{A}$	
product of matrices $\boldsymbol{A}$ and $\boldsymbol{B}$	$\boldsymbol{AB}$, where $(\boldsymbol{AB})_{ik} = \sum\limits_j A_{ij}B_{jk}$	
unit matrix	$\boldsymbol{E}$, $\boldsymbol{I}$	
inverse of a square matrix $\boldsymbol{A}$	$\boldsymbol{A}^{-1}$	
transpose of matrix $\boldsymbol{A}$	$\boldsymbol{A}^\mathsf{T}$, $\widetilde{\boldsymbol{A}}$	
complex conjugate of matrix $\boldsymbol{A}$	$\boldsymbol{A}^*$	
conjugate transpose (adjoint) of $\boldsymbol{A}$ (hermitian conjugate of $\boldsymbol{A}$)	$\boldsymbol{A}^\mathsf{H}$, $\boldsymbol{A}^\dagger$, where $\left(\boldsymbol{A}^\dagger\right)_{ij} = A_{ji}{}^*$	
trace of a square matrix $\boldsymbol{A}$	$\sum\limits_i A_{ii}$, $\mathrm{tr}\ \boldsymbol{A}$	
determinant of a square matrix $\boldsymbol{A}$	$\det \boldsymbol{A}$, $|\boldsymbol{A}|$	

sets and logical operators

Description	Symbol	Notes
p and q (conjunction sign)	$p \wedge q$	
p or q or both (disjunction sign)	$p \vee q$	
negation of p, not p	$\neg p$	
p implies q	$p \Rightarrow q$	
p is equivalent to q	$p \Leftrightarrow q$	

Description	Symbol	Notes
union of A and B	$A \cup B$	
intersection of A and B	$A \cap B$	
x belongs to A	$x \in A$	
x does not belong to A	$x \notin A$	
the set A contains x	$A \ni x$	
A but not B	$A \backslash B$	

5 FUNDAMENTAL PHYSICAL CONSTANTS

The data given in this table are from the CODATA recommended values of the fundamental physical constants 2006 [23] (online at http://physics.nist.gov/constants) and from the 2006 compilation of the Particle Data Group [133] (online at http://pdg.lbl.gov), see notes for details. The standard deviation uncertainty in the least significant digits is given in parentheses.

Quantity	Symbol	Value	Notes
magnetic constant	μ_0	$4\pi \times 10^{-7}$ H m^{-1} (defined)	1
speed of light in vacuum	c_0, c	$299\ 792\ 458$ m s^{-1} (defined)	
electric constant	$\varepsilon_0 = 1/\mu_0 c_0{}^2$	$8.854\ 187\ 817... \times 10^{-12}$ F m^{-1}	1, 2
characteristic impedance of vacuum	$Z_0 = \mu_0 c_0$	$376.730\ 313\ 461...$ Ω	2
Planck constant	h	$6.626\ 068\ 96(33) \times 10^{-34}$ J s	
	$\hbar = h/2\pi$	$1.054\ 571\ 628(53) \times 10^{-34}$ J s	
	hc_0	$1.986\ 445\ 501(99) \times 10^{-25}$ J m	
Fermi coupling constant	$G_F/(\hbar c_0)^3$	$1.166\ 37(1) \times 10^{-5}$ GeV^{-2}	3
weak mixing angle θ_W	$\sin^2 \theta_W$	$0.222\ 55(56)$	4, 5
elementary charge	e	$1.602\ 176\ 487(40) \times 10^{-19}$ C	
electron mass	m_e	$9.109\ 382\ 15(45) \times 10^{-31}$ kg	
proton mass	m_p	$1.672\ 621\ 637(83) \times 10^{-27}$ kg	
neutron mass	m_n	$1.674\ 927\ 211(84) \times 10^{-27}$ kg	
atomic mass constant	$m_u = 1$ u	$1.660\ 538\ 782(83) \times 10^{-27}$ kg	6
Avogadro constant	L, N_A	$6.022\ 141\ 79(30) \times 10^{23}$ mol^{-1}	7
Boltzmann constant	k, k_B	$1.380\ 650\ 4(24) \times 10^{-23}$ J K^{-1}	
Faraday constant	$F = Le$	$9.648\ 533\ 99(24) \times 10^4$ C mol^{-1}	
molar gas constant	R	$8.314\ 472(15)$ J K^{-1} mol^{-1}	
zero of the Celsius scale		273.15 K (defined)	
molar volume of ideal gas, $p = 100$ kPa, $t = 0$ °C	V_m	$22.710\ 981(40)$ dm^3 mol^{-1}	
molar volume of ideal gas, $p = 101.325$ kPa, $t = 0$ °C		$22.413\ 996(39)$ dm^3 mol^{-1}	
standard atmosphere		$101\ 325$ Pa (defined)	
fine-structure constant	$\alpha = \mu_0 e^2 c_0/2h$	$7.297\ 352\ 537\ 6(50) \times 10^{-3}$	
	α^{-1}	$137.035\ 999\ 676(94)$	
Bohr radius	$a_0 = 4\pi\varepsilon_0\hbar^2/m_e e^2$	$5.291\ 772\ 085\ 9(36) \times 10^{-11}$ m	
Hartree energy	$E_h = \hbar^2/m_e a_0{}^2$	$4.359\ 743\ 94(22) \times 10^{-18}$ J	
Rydberg constant	$R_\infty = E_h/2hc_0$	$1.097\ 373\ 156\ 852\ 7(73) \times 10^7$ m^{-1}	
Bohr magneton	$\mu_B = e\hbar/2m_e$	$9.274\ 009\ 15(23) \times 10^{-24}$ J T^{-1}	
electron magnetic moment	μ_e	$-9.284\ 763\ 77(23) \times 10^{-24}$ J T^{-1}	
Landé g-factor for free electron	$g_e = 2\mu_e/\mu_B$	$-2.002\ 319\ 304\ 362\ 2(15)$	
nuclear magneton	$\mu_N = e\hbar/2m_p$	$5.050\ 783\ 24(13) \times 10^{-27}$ J T^{-1}	

(1) H m^{-1} = N A^{-2} = N s^2 C^{-2}; F m^{-1} = C^2 J^{-1} m^{-1}.

(2) ε_0 and Z_0 may be calculated exactly from the defined values of μ_0 and c_0.

(3) The value of the Fermi coupling constant is recommended by the Particle Data Group [133].

(4) With the weak mixing angle θ_W, $\sin^2 \theta_W$ is sometimes called Weinberg parameter. There are a number of schemes differing in the masses used to determine $\sin^2 \theta_W$ (see Chapter 10 in [133]). The value given here for $\sin^2 \theta_W$ [23] is based on the on-shell scheme which uses $\sin^2 \theta_W = 1 - (m_W/m_Z)^2$, where the quantities m_W and m_Z are the masses of the W$^\pm$- and Z^0-bosons, respectively.

(5) The Particle Data Group [133] gives $m_W = 80.403(29)$ GeV/$c_0{}^2$, $m_Z = 91.1876(21)$ GeV/$c_0{}^2$ and recommends $\sin^2 \theta_W = 0.231\ 22(15)$, based on the $\overline{\text{MS}}$ scheme. The corresponding value in

Quantity	Symbol	Value	Notes
proton magnetic moment	μ_{p}	$1.410\ 606\ 662(37) \times 10^{-26}$ J T^{-1}	
proton gyromagnetic ratio	$\gamma_{\text{p}} = 4\pi\mu_{\text{p}}/h$	$2.675\ 222\ 099(70) \times 10^{8}$ s^{-1} T^{-1}	
shielded proton magnetic moment (H$_2$O, sphere, 25 °C)	$\mu_{\text{p}}'/\mu_{\text{B}}$	$1.520\ 993\ 128(17) \times 10^{-3}$	
shielded proton gyromagnetic ratio (H$_2$O, sphere, 25 °C)	$\gamma_{\text{p}}'/2\pi$	$42.576\ 388\ 1(12)$ MHz T^{-1}	
Stefan-Boltzmann constant	$\sigma = 2\pi^5 k^4/15h^3 c_0^2$	$5.670\ 400(40) \times 10^{-8}$ W m^{-2} K^{-4}	
first radiation constant	$c_1 = 2\pi hc_0^2$	$3.741\ 771\ 18(19) \times 10^{-16}$ W m^2	
second radiation constant	$c_2 = hc_0/k$	$1.438\ 775\ 2(25) \times 10^{-2}$ m K	
Newtonian constant of gravitation	G	$6.674\ 28(67) \times 10^{-11}$ m^3 kg^{-1} s^{-2}	
standard acceleration of gravity	g_{n}	$9.806\ 65$ m s^{-2} (defined)	

(5) (continued) the on-shell scheme is $\sin^2\theta_{\text{W}} = 0.223\ 06(33)$. The effective parameter also depends on the energy range or momentum transfer considered.

(6) u is the (unified) atomic mass unit (see Section 3.9, p. 94).

(7) See [134] and other papers in the same special issue of *Metrologia* on the precise measurement of the Avogadro constant.

Values of common mathematical constants

Mathematical constant	Symbol	Value	Notes
ratio of circumference to diameter of a circle	π	$3.141\ 592\ 653\ 59\cdots$	1
base of natural logarithms	e	$2.718\ 281\ 828\ 46\cdots$	
natural logarithm of 10	ln 10	$2.302\ 585\ 092\ 99\cdots$	

(1) A mnemonic for π, based on the number of letters in words of the English language, is:

"How I like a drink, alcoholic of course, after the heavy lectures involving quantum mechanics!"

There are similar mnemonics in poem form in French:

"Que j'aime à faire apprendre ce nombre utile aux sages!
Immortel Archimède, artiste ingénieur,
Qui de ton jugement peut priser la valeur?
Pour moi, ton problème eut de pareils avantages."

and German:

"Wie? O! Dies π
Macht ernstlich so vielen viele Müh'!
Lernt immerhin, Jünglinge, leichte Verselein,
Wie so zum Beispiel dies dürfte zu merken sein!"

See the Japanese [2.e] and Russian [2.c] editions for further mnemonics.

6 PROPERTIES OF PARTICLES, ELEMENTS, AND NUCLIDES

The symbols for particles, chemical elements, and nuclides have been discussed in Section 2.10.1 (ii), p. 50. The Particle Data Group [133] recommends the use of italic symbols for particles and this has been adopted by many physicists (see also Section 1.6, p. 7).

6.1 PROPERTIES OF SELECTED PARTICLES

The data given in this table are from the CODATA recommended values of the fundamental physical constants 2006 [23] (online at `http://physics.nist.gov/constants`) and from the 2006 compilation of the Particle Data Group (PDG) [133] (online at `http://pdg.lbl.gov`), see notes for details. The standard deviation uncertainty in the least significant digits is given in parentheses.

Name	Symbol	Spin I	Charge number z	Mass m/u	Mass mc_0^2/MeV	Notes
photon	γ	1	0	0	0	
neutrino	ν_e	1/2	0	≈ 0	≈ 0	1, 2
electron	e^-	1/2	-1	$5.485\ 799\ 094\ 3(23) \times 10^{-4}$	$0.510\ 998\ 910(13)$	3
muon	$\mu^\pm$	1/2	± 1	$0.113\ 428\ 926\ 4(30)$	$105.658\ 369\ 2(94)$	2
pion	$\pi^\pm$	0	± 1	$0.149\ 834\ 76(37)$	$139.570\ 18(35)$	2
pion	π^0	0	0	$0.144\ 903\ 35(64)$	$134.9766(6)$	2
proton	p	1/2	1	$1.007\ 276\ 466\ 77(10)$	$938.272\ 013(23)$	
neutron	n	1/2	0	$1.008\ 664\ 915\ 97(43)$	$939.565\ 346(23)$	
deuteron	d	1	1	$2.013\ 553\ 212\ 724(78)$	$1875.612\ 793(47)$	
triton	t	1/2	1	$3.015\ 500\ 713\ 4(25)$	$2808.920\ 906(70)$	4
helion	h	1/2	2	$3.014\ 932\ 247\ 3(26)$	$2808.391\ 383(70)$	4
α-particle	α	0	2	$4.001\ 506\ 179\ 127(62)$	$3727.379\ 109(93)$	
Z-boson	Z^0	1	0		$91.1876(21) \times 10^3$	2, 5
W-boson	$W^\pm$	1	± 1		$80.403(29) \times 10^3$	2, 5

(1) The neutrino and antineutrino may perhaps have a small mass, $m_{\bar{\nu}_e} < 2\ \text{eV}/c_0^2$ [133]. In addition to the electron neutrino ν_e one finds also a tau neutrino, ν_τ, and a myon neutrino, ν_μ (and their antiparticles $\bar{\nu}$).

(2) These data are from the Particle Data Group [133].

(3) The electron is sometimes denoted by e or as a β-particle by β^-. Its anti particle e^+ (positron, also β^+) has the same mass as the electron e^- but opposite charge and opposite magnetic moment.

(4) Triton is the $^3\text{H}^+$, and helion the $^3\text{He}^{2+}$ particle.

(5) Z^0 and $W^\pm$ are gauge bosons [133].

Name	Symbol	Magnetic moment μ/μ_N	Mean life [1] τ/s	Notes
photon	γ	0		
neutrino	ν_e	≈ 0		2, 6
electron	e^-	$-1.001\ 159\ 652\ 181\ 11(74)$		7, 8
muon	μ^+	$8.890\ 596\ 98(23)$	$2.197\ 03(4) \times 10^{-6}$	2, 8, 9
pion	$\pi^\pm$	0	$2.6033(5) \times 10^{-8}$	2
pion	π^0	0	$8.4(6) \times 10^{-17}$	2
proton	p	$2.792\ 847\ 356(23)$		8, 10
neutron	n	$-1.913\ 042\ 73(45)$	$885.7(8)$	8
deuteron	d	$0.857\ 438\ 230\ 8(72)$		8
triton	t	$2.978\ 962\ 448(38)$		8, 11
helion	h	$-2.127\ 497\ 718(25)$		8, 12
α-particle	α	0		

[1] The PDG [133] gives the mean life (τ) values, see also Section 2.12, note 8, p. 64.

(6) The Particle Data Group [133] gives $\mu/\mu_B < 0.9 \times 10^{-10}$.

(7) The value of the magnetic moment is given in Bohr magnetons μ/μ_B, $\mu_B = e\hbar/2m_e$.

In nuclear physics and chemistry the masses of particles are often quoted as their energy equivalents (usually in megaelectronvolts). The unified atomic mass unit corresponds to 931.494 028(23) MeV [23].

Atom-like pairs of a positive particle and an electron are sometimes sufficiently stable to be treated as individual entities with special names.

Examples   positronium (e^+e^-; Ps) $m(\text{Ps}) = 1.097\ 152\ 515\ 21(46) \times 10^{-3}$ u

muonium (μ^+e^-; Mu) $m(\text{Mu}) = 0.113\ 977\ 490\ 9(29)$ u

---

(Notes continued)

(8) The sign of the magnetic moment is defined with respect to the direction of the spin angular momentum.

(9) μ^- and μ^+ have the same mass but opposite charge and opposite magnetic moment.

(10) The shielded proton magnetic moment, μ_p', is given by $\mu_p'/\mu_N = 2.792\ 775\ 598(30)$ (H_2O, sphere, 25 °C).

(11) The half life, $t_{1/2}$, of the triton is about 12.3 a (see Section 2.12, p. 64) with a corresponding mean life, τ, of 17.7 a.

(12) This is the shielded helion magnetic moment, μ_h', given as μ_h'/μ_N (gas, sphere, 25 °C).

116

6.2 STANDARD ATOMIC WEIGHTS OF THE ELEMENTS 2005

As agreed by the IUPAC Commission on Atomic Weights and Isotopic Abundances (CAWIA) in 1979 [135] the relative atomic mass (generally called atomic weight [136]) of an element, E, can be defined for any specified sample. It is the average mass of the atoms in the sample divided by the unified atomic mass unit[1] or alternatively the molar mass of its atoms divided by the molar mass constant $M_u = N_A m_u = 1$ g mol^{-1}:

$$A_r(E) = \overline{m}_a(E)/u = M(E)/M_u$$

The variations in isotopic composition of many elements in samples of different origin limit the precision to which an atomic weight can be given. The standard atomic weights revised biennially by the CAWIA are meant to be applicable for normal materials. This means that to a high level of confidence the atomic weight of an element in any normal sample will be within the uncertainty limits of the tabulated value. By "normal" it is meant here that the material is a reasonably possible source of the element or its compounds in commerce for industry and science and that it has not been subject to significant modification of isotopic composition within a geologically brief period [137]. This, of course, excludes materials studied themselves for very anomalous isotopic composition. New statistical guidelines have been formulated and used to provide uncertainties on isotopic abundances in the isotopic composition of the elements 1997 [138].

Table 6.2 below lists the atomic weights of the elements 2005 [139] and the term symbol $^{2S+1}L_J$ for the atomic ground state [140] in the order of the atomic number. The atomic weights have been recommended by the IUPAC Commission on Isotopic Abundances and Atomic Weights (CIAAW) in 2005 [139] and apply to elements as they exist naturally on earth. An electronic version of the CIAAW Table of Standard Atomic Weights 2005 can be found on the CIAAW web page at http://www.ciaaw.org/atomic_weights4.htm. The list includes the approved names of elements 110 and 111 (Ds and Rg) [141,142]. The symbol Rg has also been used for "rare gas". For a history of recommended atomic weight values from 1882 to 1997, see [143].

The atomic weights of many elements depend on the origin and treatment of the materials [138]. The notes to this table explain the types of variation to be expected for individual elements. When used with due regard to the notes the values are considered reliable to ± the figure given in parentheses being applicable to the last digit. For elements without a characteristic terrestrial isotopic composition no standard atomic weight is recommended. The atomic mass of its most stable isotope can be found in Section 6.3 below.

Symbol	Atomic number	Name	Atomic weight (Relative atomic mass)	Ground state term symbol	Note
H	1	hydrogen	1.007 94(7)	$^2S_{1/2}$	g, m, r
He	2	helium	4.002 602(2)	$^1S_0$	g, r
Li	3	lithium	[6.941(2)]†	$^2S_{1/2}$	g, m, r
Be	4	beryllium	9.012 182(3)	$^1S_0$	
B	5	boron	10.811(7)	$^2P^o_{1/2}$	g, m, r
C	6	carbon	12.0107(8)	$^3P_0$	g, r
N	7	nitrogen	14.0067(2)	$^4S^o_{3/2}$	g, r
O	8	oxygen	15.9994(3)	$^3P_2$	g, r
F	9	fluorine	18.998 403 2(5)	$^2P^o_{3/2}$	
Ne	10	neon	20.1797(6)	$^1S_0$	g, m
Na	11	sodium	22.989 769 28(2)	$^2S_{1/2}$	
Mg	12	magnesium	24.3050(6)	$^1S_0$	

[1] Note that the atomic mass constant m_u is equal to the Dalton, Da, or the unified atomic mass unit, u, and is defined in terms of the mass of the carbon-12 atom: $m_u = 1$ u $= 1$ Da $= m_a(^{12}C)/12$.

Symbol	Atomic number	Name	Atomic weight (Relative atomic mass)	Ground state term symbol	Note
Al	13	aluminium (aluminum)	26.981 538 6(8)	$^2P^o_{1/2}$	
Si	14	silicon	28.0855(3)	$^3P_0$	r
P	15	phosphorus	30.973 762(2)	$^4S^o_{3/2}$	
S	16	sulfur	32.065(5)	$^3P_2$	g, r
Cl	17	chlorine	35.453(2)	$^2P^o_{3/2}$	g, m, r
Ar	18	argon	39.948(1)	$^1S_0$	g, r
K	19	potassium	39.0983(1)	$^2S_{1/2}$	
Ca	20	calcium	40.078(4)	$^1S_0$	g
Sc	21	scandium	44.955 912(6)	$^2D_{3/2}$	
Ti	22	titanium	47.867(1)	$^3F_2$	
V	23	vanadium	50.9415(1)	$^4F_{3/2}$	
Cr	24	chromium	51.9961(6)	$^7S_3$	
Mn	25	manganese	54.938 045(5)	$^6S_{5/2}$	
Fe	26	iron	55.845(2)	$^5D_4$	
Co	27	cobalt	58.933 195(5)	$^4F_{9/2}$	
Ni	28	nickel	58.6934(2)	$^3F_4$	
Cu	29	copper	63.546(3)	$^2S_{1/2}$	r
Zn	30	zinc	65.409(4)	$^1S_0$	
Ga	31	gallium	69.723(1)	$^2P^o_{1/2}$	
Ge	32	germanium	72.64(1)	$^3P_0$	
As	33	arsenic	74.921 60(2)	$^4S^o_{3/2}$	
Se	34	selenium	78.96(3)	$^3P_2$	r
Br	35	bromine	79.904(1)	$^2P^o_{3/2}$	
Kr	36	krypton	83.798(2)	$^1S_0$	g, m
Rb	37	rubidium	85.4678(3)	$^2S_{1/2}$	g
Sr	38	strontium	87.62(1)	$^1S_0$	g, r
Y	39	yttrium	88.905 85(2)	$^2D_{3/2}$	
Zr	40	zirconium	91.224(2)	$^3F_2$	g
Nb	41	niobium	92.906 38(2)	$^6D_{1/2}$	
Mo	42	molybdenum	95.94(2)	$^7S_3$	g
Tc	43	technetium		$^6S_{5/2}$	A
Ru	44	ruthenium	101.07(2)	$^5F_5$	g
Rh	45	rhodium	102.905 50(2)	$^4F_{9/2}$	
Pd	46	palladium	106.42(1)	$^1S_0$	g
Ag	47	silver	107.8682(2)	$^2S_{1/2}$	g
Cd	48	cadmium	112.411(8)	$^1S_0$	g
In	49	indium	114.818(3)	$^2P^o_{1/2}$	
Sn	50	tin	118.710(7)	$^3P_0$	g
Sb	51	antimony	121.760(1)	$^4S^o_{3/2}$	g
Te	52	tellurium	127.60(3)	$^3P_2$	g
I	53	iodine	126.904 47(3)	$^2P^o_{3/2}$	
Xe	54	xenon	131.293(6)	$^1S_0$	g, m
Cs	55	caesium (cesium)	132.905 451 9(2)	$^2S_{1/2}$	
Ba	56	barium	137.327(7)	$^1S_0$	
La	57	lanthanum	138.905 47(7)	$^2D_{3/2}$	g
Ce	58	cerium	140.116(1)	$^1G^o_4$	g

Symbol	Atomic number	Name	Atomic weight (Relative atomic mass)	Ground state term symbol	Note
Pr	59	praseodymium	140.907 65(2)	$^4I^o_{9/2}$	
Nd	60	neodymium	144.242(3)	$^5I_4$	g
Pm	61	promethium		$^6H^o_{5/2}$	A
Sm	62	samarium	150.36(2)	$^7F_0$	g
Eu	63	europium	151.964(1)	$^8S^o_{7/2}$	g
Gd	64	gadolinium	157.25(3)	$^9D^o_2$	g
Tb	65	terbium	158.925 35(2)	$^6H^o_{15/2}$	
Dy	66	dysprosium	162.500(1)	$^5I_8$	g
Ho	67	holmium	164.930 32(2)	$^4I^o_{15/2}$	
Er	68	erbium	167.259(3)	$^3H_6$	g
Tm	69	thulium	168.934 21(2)	$^2F^o_{7/2}$	
Yb	70	ytterbium	173.04(3)	$^1S_0$	g
Lu	71	lutetium	174.967(1)	$^2D_{3/2}$	g
Hf	72	hafnium	178.49(2)	$^3F_2$	
Ta	73	tantalum	180.947 88(2)	$^4F_{3/2}$	
W	74	tungsten	183.84(1)	$^5D_0$	
Re	75	rhenium	186.207(1)	$^6S_{5/2}$	
Os	76	osmium	190.23(3)	$^5D_4$	g
Ir	77	iridium	192.217(3)	$^4F_{9/2}$	
Pt	78	platinum	195.084(9)	$^3D_3$	
Au	79	gold	196.966 569(4)	$^2S_{1/2}$	
Hg	80	mercury	200.59(2)	$^1S_0$	
Tl	81	thallium	204.3833(2)	$^2P^o_{1/2}$	
Pb	82	lead	207.2(1)	$^3P_0$	g, r
Bi	83	bismuth	208.980 40(1)	$^4S^o_{3/2}$	
Po	84	polonium		$^3P_2$	A
At	85	astatine		$^2P^o_{3/2}$	A
Rn	86	radon		$^1S_0$	A
Fr	87	francium		$^2S_{1/2}$	A
Ra	88	radium		$^1S_0$	A
Ac	89	actinium		$^2D_{3/2}$	A
Th	90	thorium	232.038 06(2)	$^3F_2$	g, Z
Pa	91	protactinium	231.035 88(2)	$^4K_{11/2}$	Z
U	92	uranium	238.028 91(3)	$^5L_6$	g, m, Z
Np	93	neptunium		$^6L_{11/2}$	A
Pu	94	plutonium		$^7F_0$	A
Am	95	americium		$^8S^o_{7/2}$	A
Cm	96	curium		$^9D^o_2$	A
Bk	97	berkelium		$^6H^o_{15/2}$	A
Cf	98	californium		$^5I_8$	A
Es	99	einsteinium		$^4I^o_{15/2}$	A
Fm	100	fermium		$^3H_6$	A
Md	101	mendelevium		$^2F^o_{7/2}$	A
No	102	nobelium		$^1S_0$	A

Symbol	Atomic number	Name	Atomic weight (Relative atomic mass)	Ground state term symbol	Note
Lr	103	lawrencium			A
Rf	104	rutherfordium			A
Db	105	dubnium			A
Sg	106	seaborgium			A
Bh	107	bohrium			A
Hs	108	hassium			A
Mt	109	meitnerium			A
Ds	110	darmstadtium			A
Rg	111	roentgenium			A

† Commercially available Li materials have atomic weights that range between 6.939 and 6.996; if a more accurate value is required, it must be determined for the specific material.

(g) **G**eological specimens are known in which the element has an isotopic composition outside the limits for normal material. The difference between the atomic weight of the element in such specimens and that given in the table may exceed the stated uncertainty.

(m) **M**odified isotopic compositions may be found in commercially available material because it has been subjected to an undisclosed or inadvertent isotopic fractionation. Substantial deviations in atomic weight of the element from that given in the table can occur.

(r) **R**ange in isotopic composition of normal terrestrial material prevents a more precise $A_r(E)$ being given; the tabulated $A_r(E)$ value and uncertainty should be applicable to normal material.

(A) Radioactive element without stable nuclide that lacks a characteristic terrestrial isotopic composition. In the IUPAC Periodic Table of the Elements on the inside back cover, a value in brackets indicates the mass number of the longest-lived isotope of the element (see also the table of nuclide masses, Section 6.3).

(Z) An element without stable nuclide(s), exhibiting a range of characteristic terrestrial compositions of long-lived radionuclide(s) such that a meaningful atomic weight can be given.

6.3 PROPERTIES OF NUCLIDES

The table contains the following properties of naturally occurring and some unstable nuclides:

Column

1   Z is the atomic number (number of protons) of the nuclide.

2   Symbol of the element.

3   A is the mass number of the nuclide. The asterisk * denotes an unstable nuclide (for elements without naturally occurring isotopes it is the most stable nuclide) and the # sign a nuclide of sufficiently long lifetime (greater than 10^5 years) [144] to enable the determination of its isotopic abundance.

4   The atomic mass is given in unified atomic mass units, 1 u $= m_\mathrm{a}(^{12}\mathrm{C})/12$, together with the standard errors in parentheses and applicable to the last digits quoted. The data were extracted from a more extensive list of the AME2003 atomic mass evaluation [145, 146].

5   Representative isotopic compositions are given as amount-of-substance fractions (mole fractions), x, of the corresponding atoms in percents. According to the opinion of CAWIA, they represent the isotopic composition of chemicals or materials most commonly encountered in the laboratory. They may not, therefore, correspond to the most abundant natural material [138]. It must be stressed that those values should be used to determine the average properties of chemicals or materials of unspecified natural terrestrial origin, though no actual sample having the exact composition listed may be available. The values listed here are from the 2001 CAWIA review as given in column 9 of ref. [147] as representative isotopic composition. This reference uses the *Atomic Mass Evaluation 1993* [148, 149]. There is an inconsistency in this Table because column 4 uses the most recent masses AME2003 [145,146], whereas column 5 is based on previous atomic masses [148,149]. When precise work is to be undertaken, such as assessment of individual properties, samples with more precisely known isotopic abundances (such as listed in column 8 of ref. [147]) should be obtained or suitable measurements should be made. The uncertainties given in parentheses are applicable to the last digits quoted and cover the range of probable variations in the materials as well as experimental errors. For additional data and background information on ranges of isotope-abundance variations in natural and anthropogenic material, see [149,150].

6   I is the nuclear spin quantum number. A plus sign indicates positive parity and a minus sign indicates negative parity. Parentheses denotes uncertain values; all values have been taken from the NUBASE evaluation [144].

7   Under magnetic moment the maximum z-component expectation value of the magnetic dipole moment, m, in nuclear magnetons is given. The positive or negative sign implies that the orientation of the magnetic dipole with respect to the angular momentum corresponds to the rotation of a positive or negative charge, respectively. The data were extracted from the compilation by N. J. Stone [151]. An asterisk * indicates that more than one value is given in the original compilation; ** indicates that an older value exists with a higher stated precision. The absence of a plus or minus sign means that the sign has not been determined by the experimenter.

8   Under quadrupole moment, the electric quadrupole moment area (see Section 2.5, notes 14 and 15 on p. 23 and 24) is given in units of square femtometres, 1 $\mathrm{fm}^2 = 10^{-30}$ m^2, although most of the tables quote them in barn (b), 1 b $= 10^{-28}$ $\mathrm{m}^2 = 100$ fm^2. The positive sign implies a prolate nucleus, the negative sign an oblate nucleus. The data are taken from N. J. Stone [151]. An asterisk * indicates that more than one value is given in the original compilation; ** indicates that an older value exists with a higher stated precision. The absence of a plus or minus sign means that the sign has not been determined by the experimenter.

Z	Symbol	A	Atomic mass, m_a/u	Isotopic composition, $100\ x$	Nuclear spin, I	Magnetic moment, m/μ_N	Quadrupole moment, Q/fm^2
1	H	1	1.007 825 032 07(10)	99.9885(70)	1/2+	+2.792 847 34(3)	
	(D)	2	2.014 101 777 8(4)	0.0115(70)	1+	+0.857 438 228(9)	+0.286(2)*
	(T)	3*	3.016 049 277 7(25)		1/2+	+2.978 962 44(4)	
2	He	3	3.016 029 319 1(26)	0.000 134(3)	1/2+	−2.127 497 72(3)	
		4	4.002 603 254 15(6)	99.999 866(3)	0+	0	
3	Li	6	6.015 122 795(16)	7.59(4)	1+	+0.822 047 3(6)*	−0.082(2)*
		7	7.016 004 55(8)	92.41(4)	3/2−	+3.256 427(2)*	−4.06(8)*
4	Be	9	9.012 182 2(4)	100	3/2−	−1.177 432(3)*	+5.29(4)*
5	B	10	10.012 937 0(4)	19.9(7)	3+	+1.800 644 78(6)	+8.47(6)
		11	11.009 305 4(4)	80.1(7)	3/2−	+2.688 648 9(10)	+4.07(3)
6	C	12	12 (by definition)	98.93(8)	0+	0	
		13	13.003 354 837 8(10)	1.07(8)	1/2−	+0.702 411 8(14)	
		14*	14.003 241 989(4)		0+	0	
7	N	14	14.003 074 004 8(6)	99.636(20)	1+	+0.403 761 00(6)	+2.001(10)*
		15	15.000 108 898 2(7)	0.364(20)	1/2−	−0.283 188 84(5)	
8	O	16	15.994 914 619 56(16)	99.757(16)	0+	0	
		17	16.999 131 70(12)	0.038(1)	5/2+	−1.893 79(9)	−2.578*
		18	17.999 161 0(7)	0.205(14)	0+	0	
9	F	19	18.998 403 22(7)	100	1/2+	+2.628 868(8)	
10	Ne	20	19.992 440 175 4(19)	90.48(3)	0+	0	
		21	20.993 846 68(4)	0.27(1)	3/2+	−0.661 797(5)	+10.3(8)
		22	21.991 385 114(19)	9.25(3)	0+	0	
11	Na	23	22.989 769 280 9(29)	100	3/2+	+2.217 655 6(6)*	+10.45(10)*
12	Mg	24	23.985 041 700(14)	78.99(4)	0+	0	
		25	24.985 836 92(3)	10.00(1)	5/2+	−0.855 45(8)	+19.9(2)*
		26	25.982 592 929(30)	11.01(3)	0+	0	
13	Al	27	26.981 538 63(12)	100	5/2+	+3.641 506 9(7)	+14.66(10)*
14	Si	28	27.976 926 532 5(19)	92.223(19)	0+	0	
		29	28.976 494 700(22)	4.685(8)	1/2+	−0.555 29(3)	
		30	29.973 770 17(3)	3.092(11)	0+	0	
15	P	31	30.973 761 63(20)	100	1/2+	+1.131 60(3)	
16	S	32	31.972 071 00(15)	94.99(26)	0+	0	
		33	32.971 458 76(15)	0.75(2)	3/2+	+0.643 821 2(14)	−6.4(10)*
		34	33.967 866 90(12)	4.25(24)	0+	0	
		36	35.967 080 76(20)	0.01(1)	0+	0	
17	Cl	35	34.968 852 68(4)	75.76(10)	3/2+	+0.821 874 3(4)	8.50(11)*
		37	36.965 902 59(5)	24.24(10)	3/2+	+0.684 123 6(4)	−6.44(7)*
18	Ar	36	35.967 545 106(29)	0.3365(30)	0+	0	
		38	37.962 732 4(4)	0.0632(5)	0+	0	
		40	39.962 383 122 5(29)	99.6003(30)	0+	0	
19	K	39	38.963 706 68(20)	93.2581(44)	3/2+	+0.391 47(3)**	+5.85
		40#	39.963 998 48(21)	0.0117(1)	4−	−1.298 100(3)	−7.3(1)*
		41	40.961 825 76(21)	6.7302(44)	3/2+	+0.214 870 1(2)**	+7.11(7)*

Z	Symbol	A	Atomic mass, m_a/u	Isotopic composition, $100\,x$	Nuclear spin, I	Magnetic moment, m/μ_N	Quadrupole moment, Q/fm^2
20	Ca	40	39.962 590 98(22)	96.941(156)	0+	0	
		42	41.958 618 01(27)	0.647(23)	0+	0	
		43	42.958 766 6(3)	0.135(10)	7/2−	−1.317 643(7)*	−5.5(1)**
		44	43.955 481 8(4)	2.086(110)	0+	0	
		46	45.953 692 6(24)	0.004(3)	0+	0	
		48#	47.952 534(4)	0.187(21)	0+	0	
21	Sc	45	44.955 911 9(9)	100	7/2−	+4.756 487(2)	−15.6(3)*
22	Ti	46	45.952 631 6(9)	8.25(3)	0+	0	
		47	46.951 763 1(9)	7.44(2)	5/2−	−0.788 48(1)	+30.0(20)*
		48	47.947 946 3(9)	73.72(3)	0+	0	
		49	48.947 870 0(9)	5.41(2)	7/2−	−1.104 17(1)	24.7(11)*
		50	49.944 791 2(9)	5.18(2)	0+	0	
23	V	50#	49.947 158 5(11)	0.250(4)	6+	+3.345 688 9(14)	21.0(40)*
		51	50.943 959 5(11)	99.750(4)	7/2−	+5.148 705 7(2)	−4.3(5)*
24	Cr	50	49.946 044 2(11)	4.345(13)	0+	0	
		52	51.940 507 5(8)	83.789(18)	0+	0	
		53	52.940 649 4(8)	9.501(17)	3/2−	−0.474 54(3)	−15.0(50)*
		54	53.938 880 4(8)	2.365(7)	0+	0	
25	Mn	55	54.938 045 1(7)	100	5/2−	+3.468 717 90(9)	+33.0(10)*
26	Fe	54	53.939 610 5(7)	5.845(35)	0+	0	
		56	55.934 937 5(7)	91.754(36)	0+	0	
		57	56.935 394 0(7)	2.119(10)	1/2−	+0.090 623 00(9)*	
		58	57.933 275 6(8)	0.282(4)	0+	0	
27	Co	59	58.933 195 0(7)	100	7/2−	+4.627(9)	+41.0(10)*
28	Ni	58	57.935 342 9(7)	68.0769(89)	0+	0	
		60	59.930 786 4(7)	26.2231(77)	0+	0	
		61	60.931 056 0(7)	1.1399(6)	3/2−	−0.750 02(4)	+16.2(15)
		62	61.928 345 1(6)	3.6345(17)	0+	0	
		64	63.927 966 0(7)	0.9256(9)	0+	0	
29	Cu	63	62.929 597 5(6)	69.15(15)	3/2−	2.227 345 6(14)*	−21.1(4)*
		65	64.927 789 5(7)	30.85(15)	3/2−	2.381 61(19)*	−19.5(4)
30	Zn	64	63.929 142 2(7)	48.268(321)	0+	0	
		66	65.926 033 4(10)	27.975(77)	0+	0	
		67	66.927 127 3(10)	4.102(21)	5/2−	+0.875 204 9(11)*	+15.0(15)
		68	67.924 844 2(10)	19.024(123)	0+	0	
		70	69.925 319 3(21)	0.631(9)	0+	0	
31	Ga	69	68.925 573 6(13)	60.108(9)	3/2−	+2.016 59(5)	+16.50(8)*
		71	70.924 701 3(11)	39.892(9)	3/2−	+2.562 27(2)	+10.40(8)*
32	Ge	70	69.924 247 4(11)	20.38(18)	0+	0	
		72	71.922 075 8(18)	27.31(26)	0+	0	
		73	72.923 458 9(18)	7.76(8)	9/2+	−0.879 467 7(2)	−17.0(30)
		74	73.921 177 8(18)	36.72(15)	0+	0	
		76#	75.921 402 6(18)	7.83(7)	0+	0	
33	As	75	74.921 596 5(20)	100	3/2−	+1.439 48(7)	+30.0(50)**

Z	Symbol	A	Atomic mass, m_a/u	Isotopic composition, $100\,x$	Nuclear spin, I	Magnetic moment, m/μ_N	Quadrupole moment, Q/fm^2
34	Se	74	73.922 476 4(18)	0.89(4)	0+	0	
		76	75.919 213 6(18)	9.37(29)	0+	0	
		77	76.919 914 0(18)	7.63(16)	1/2−	+0.535 042 2(6)*	
		78	77.917 309 1(18)	23.77(28)	0+	0	
		80	79.916 521 3(21)	49.61(41)	0+	0	
		82#	81.916 699 4(22)	8.73(22)	0+	0	
35	Br	79	78.918 337 1(22)	50.69(7)	3/2−	+2.106 400(4)	31.8(5)**
		81	80.916 290 6(21)	49.31(7)	3/2−	+2.270 562(4)	+26.6(4)**
36	Kr	78	77.920 364 8(12)	0.355(3)	0+	0	
		80	79.916 379 0(16)	2.286(10)	0+	0	
		82	81.913 483 6(19)	11.593(31)	0+	0	
		83	82.914 136(3)	11.500(19)	9/2+	−0.970 669(3)	+25.9(1)*
		84	83.911 507(3)	56.987(15)	0+	0	
		86	85.910 610 73(11)	17.279(41)	0+	0	
37	Rb	85	84.911 789 738(12)	72.17(2)	5/2−	+1.352 98(10)**	+27.7(1)**
		87#	86.909 180 527(13)	27.83(2)	3/2−	+2.751 31(12)**	+13.4(1)*
38	Sr	84	83.913 425(3)	0.56(1)	0+	0	
		86	85.909 260 2(12)	9.86(1)	0+	0	
		87	86.908 877 1(12)	7.00(1)	9/2+	−1.093 603 0(13)*	+33.0(20)*
		88	87.905 612 1(12)	82.58(1)	0+	0	
39	Y	89	88.905 848 3(27)	100	1/2−	−0.137 415 4(3)*	
40	Zr	90	89.904 704 4(25)	51.45(40)	0+	0	
		91	90.905 645 8(25)	11.22(5)	5/2+	−1.303 62(2)	−17.6(3)*
		92	91.905 040 8(25)	17.15(8)	0+	0	
		94	93.906 315 2(26)	17.38(28)	0+	0	
		96#	95.908 273 4(30)	2.80(9)	0+	0	
41	Nb	93	92.906 378 1(26)	100	9/2+	+6.1705(3)	−37.0(20)*
42	Mo	92	91.906 811(4)	14.77(31)	0+	0	
		94	93.905 088 3(21)	9.23(10)	0+	0	
		95	94.905 842 1(21)	15.90(9)	5/2+	−0.9142(1)	−2.2(1)*
		96	95.904 679 5(21)	16.68(1)	0+	0	
		97	96.906 021 5(21)	9.56(5)	5/2+	−0.9335(1)	+25.5(13)*
		98	97.905 408 2(21)	24.19(26)	0+	0	
		100#	99.907 477(6)	9.67(20)	0+	0	
43	Tc	98*	97.907 216(4)		(6)+		
44	Ru	96	95.907 598(8)	5.54(14)	0+	0	
		98	97.905 287(7)	1.87(3)	0+	0	
		99	98.905 939 3(22)	12.76(14)	5/2+	−0.641(5)	+7.9(4)
		100	99.904 219 5(22)	12.60(7)	0+	0	
		101	100.905 582 1(22)	17.06(2)	5/2+	−0.719(6)*	+46.0(20)
		102	101.904 349 3(22)	31.55(14)	0+	0	
		104	103.905 433(3)	18.62(27)	0+	0	
45	Rh	103	102.905 504(3)	100	1/2−	−0.8840(2)	
46	Pd	102	101.905 609(3)	1.02(1)	0+	0	
		104	103.904 036(4)	11.14(8)	0+	0	
		105	104.905 085(4)	22.33(8)	5/2+	−0.642(3)	+65.0(30)**
		106	105.903 486(4)	27.33(3)	0+	0	
		108	107.903 892(4)	26.46(9)	0+	0	
		110	109.905 153(12)	11.72(9)	0+	0	

Z	Symbol	A	Atomic mass, m_a/u	Isotopic composition, $100\,x$	Nuclear spin, I	Magnetic moment, m/μ_N	Quadrupole moment, Q/fm^2
47	Ag	107	106.905 097(5)	51.839(8)	1/2−	−0.113 679 65(15)*	
		109	108.904 752(3)	48.161(8)	1/2−	−0.130 690 6(2)*	
48	Cd	106	105.906 459(6)	1.25(6)	0+	0	
		108	107.904 184(6)	0.89(3)	0+	0	
		110	109.903 002 1(29)	12.49(18)	0+	0	
		111	110.904 178 1(29)	12.80(12)	1/2+	−0.594 886 1(8)*	
		112	111.902 757 8(29)	24.13(21)	0+	0	
		113#	112.904 401 7(29)	12.22(12)	1/2+	−0.622 300 9(9)	
		114	113.903 358 5(29)	28.73(42)	0+	0	
		116#	115.904 756(3)	7.49(18)	0+	0	
49	In	113	112.904 058(3)	4.29(5)	9/2+	+5.5289(2)	+80.0(40)
		115#	114.903 878(5)	95.71(5)	9/2+	+5.5408(2)	+81.0(50)*
50	Sn	112	111.904 818(5)	0.97(1)	0+	0	
		114	113.902 779(3)	0.66(1)	0+	0	
		115	114.903 342(3)	0.34(1)	1/2+	−0.918 83(7)	
		116	115.901 741(3)	14.54(9)	0+	0	
		117	116.902 952(3)	7.68(7)	1/2+	−1.001 04(7)	
		118	117.901 603(3)	24.22(9)	0+	0	
		119	118.903 308(3)	8.59(4)	1/2+	−1.047 28(7)	
		120	119.902 194 7(27)	32.58(9)	0+	0	
		122	121.903 439 0(29)	4.63(3)	0+	0	
		124	123.905 273 9(15)	5.79(5)	0+	0	
51	Sb	121	120.903 815 7(24)	57.21(5)	5/2+	+3.3634(3)	−36.0(40)**
		123	122.904 214 0(22)	42.79(5)	7/2+	+2.5498(2)	−49.0(50)
52	Te	120	119.904 020(10)	0.09(1)	0+	0	
		122	121.903 043 9(16)	2.55(12)	0+	0	
		123#	122.904 270 0(16)	0.89(3)	1/2+	−0.736 947 8(8)	
		124	123.902 817 9(16)	4.74(14)	0+	0	
		125	124.904 430 7(16)	7.07(15)	1/2+	−0.888 450 9(10)*	
		126	125.903 311 7(16)	18.84(25)	0+	0	
		128#	127.904 463 1(19)	31.74(8)	0+	0	
		130#	129.906 224 4(21)	34.08(62)	0+	0	
53	I	127	126.904 473(4)	100	5/2+	+2.813 27(8)	72.0(20)**
54	Xe	124	123.905 893 0(20)	0.0952(3)	0+	0	
		126	125.904 274(7)	0.0890(2)	0+	0	
		128	127.903 531 3(15)	1.9102(8)	0+	0	
		129	128.904 779 4(8)	26.4006(82)	1/2+	−0.777 976(8)	
		130	129.903 508 0(8)	4.0710(13)	0+	0	
		131	130.905 082 4(10)	21.2324(30)	3/2+	+0.6915(2)**	−11.4(1)*
		132	131.904 153 5(10)	26.9086(33)	0+	0	
		134	133.905 394 5(9)	10.4357(21)	0+	0	
		136	135.907 219(8)	8.8573(44)	0+	0	
55	Cs	133	132.905 451 933(24)	100	7/2+	+2.582 025(3)**	−0.355(4)*
56	Ba	130	129.906 320(8)	0.106(1)	0+	0	
		132	131.905 061(3)	0.101(1)	0+	0	
		134	133.904 508 4(4)	2.417(18)	0+	0	
		135	134.905 688 6(4)	6.592(12)	3/2+	0.838 627(2)*	+16.0(3)*
		136	135.904 575 9(4)	7.854(24)	0+	0	
		137	136.905 827 4(5)	11.232(24)	3/2+	0.937 34(2)*	+24.5(4)*
		138	137.905 247 2(5)	71.698(42)	0+	0	

125

Z	Symbol	A	Atomic mass, m_a/u	Isotopic composition, $100\,x$	Nuclear spin, I	Magnetic moment, m/μ_N	Quadrupole moment, Q/fm^2
57	La	138#	137.907 112(4)	0.090(1)	5+	+3.713 646(7)	+45.0(20)*
		139	138.906 353 3(26)	99.910(1)	7/2+	+2.783 045 5(9)	+20.0(10)
58	Ce	136	135.907 172(14)	0.185(2)	0+	0	
		138	137.905 991(11)	0.251(2)	0+	0	
		140	139.905 438 7(26)	88.450(51)	0+	0	
		142	141.909 244(3)	11.114(51)	0+	0	
59	Pr	141	140.907 652 8(26)	100	5/2+	+4.2754(5)	−7.7(6)*
60	Nd	142	141.907 723 3(25)	27.2(5)	0+	0	
		143	142.909 814 3(25)	12.2(2)	7/2−	−1.065(5)	−61.0(20)*
		144#	143.910 087 3(25)	23.8(3)	0+	0	
		145	144.912 573 6(25)	8.3(1)	7/2−	−0.656(4)	−31.4(12)**
		146	145.913 116 9(25)	17.2(3)	0+	0	
		148	147.916 893(3)	5.7(1)	0+	0	
		150#	149.920 891(3)	5.6(2)	0+	0	
61	Pm	145*	144.912 749(3)		5/2+		
62	Sm	144	143.911 999(3)	3.07(7)	0+	0	
		147#	146.914 897 9(26)	14.99(18)	7/2−	−0.812(2)**	−26.1(7)*
		148#	147.914 822 7(26)	11.24(10)	0+	0	
		149	148.917 184 7(26)	13.82(7)	7/2−	−0.6677(11)**	+7.5(2)*
		150	149.917 275 5(26)	7.38(1)	0+	0	
		152	151.919 732 4(27)	26.75(16)	0+	0	
		154	153.922 209 3(27)	22.75(29)	0+	0	
63	Eu	151	150.919 850 2(26)	47.81(6)	5/2+	+3.4717(6)	83.0**
		153	152.921 230 3(26)	52.19(6)	5/2+	+1.5324(3)*	+222.0*
64	Gd	152#	151.919 791 0(27)	0.20(1)	0+	0	
		154	153.920 865 6(27)	2.18(3)	0+	0	
		155	154.922 622 0(27)	14.80(12)	3/2−	−0.2572(4)*	+127.0(50)*
		156	155.922 122 7(27)	20.47(9)	0+	0	
		157	156.923 960 1(27)	15.65(2)	3/2−	−0.3373(6)*	+136.0(60)**
		158	157.924 103 9(27)	24.84(7)	0+	0	
		160	159.927 054 1(27)	21.86(19)	0+	0	
65	Tb	159	158.925 346 8(27)	100	3/2+	+2.014(4)	+143.2(8)
66	Dy	156	155.924 283(7)	0.056(3)	0+	0	
		158	157.924 409(4)	0.095(3)	0+	0	
		160	159.925 197 5(27)	2.329(18)	0+	0	
		161	160.926 933 4(27)	18.889(42)	5/2+	−0.480(3)*	247.7(30)**
		162	161.926 798 4(27)	25.475(36)	0+	0	
		163	162.928 731 2(27)	24.896(42)	5/2−	+0.673(4)	+265.0(20)**
		164	163.929 174 8(27)	28.260(54)	0+	0	
67	Ho	165	164.930 322 1(27)	100	7/2−	+4.17(3)	358.0(20)**
68	Er	162	161.928 778(4)	0.139(5)	0+	0	
		164	163.929 200(3)	1.601(3)	0+	0	
		166	165.930 293 1(27)	33.503(36)	0+	0	
		167	166.932 048 2(27)	22.869(9)	7/2+	−0.563 85(12)	+357.0(3)**
		168	167.932 370 2(27)	26.978(18)	0+	0	
		170	169.935 464 3(30)	14.910(36)	0+	0	
69	Tm	169	168.934 213 3(27)	100	1/2+	−0.2310(15)*	

Z	Symbol	A	Atomic mass, m_a/u	Isotopic composition, $100\ x$	Nuclear spin, I	Magnetic moment, m/μ_N	Quadrupole moment, Q/fm^2
70	Yb	168	167.933 897(5)	0.13(1)	0+	0	
		170	169.934 761 8(26)	3.04(15)	0+	0	
		171	170.936 325 8(26)	14.28(57)	1/2−	+0.493 67(1)*	
		172	171.936 381 5(26)	21.83(67)	0+	0	
		173	172.938 210 8(26)	16.13(27)	5/2−	−0.648(3)**	+280.0(40)
		174	173.938 862 1(26)	31.83(92)	0+	0	
		176	175.942 571 7(28)	12.76(41)	0+	0	
71	Lu	175	174.940 771 8(23)	97.41(2)	7/2+	+2.2323(11)**	+349.0(20)*
		176#	175.942 686 3(23)	2.59(2)	7−	+3.162(12)**	+492.0(50)*
72	Hf	174#	173.940 046(3)	0.16(1)	0+	0	
		176	175.941 408 6(24)	5.26(7)	0+	0	
		177	176.943 220 7(23)	18.60(9)	7/2−	+0.7935(6)	+337.0(30)*
		178	177.943 698 8(23)	27.28(7)	0+	0	
		179	178.945 816 1(23)	13.62(2)	9/2+	−0.6409(13)	+379.0(30)*
		180	179.946 550 0(23)	35.08(16)	0+	0	
73	Ta	180	179.947 464 8(24)	0.012(2)	9−		
		181	180.947 995 8(19)	99.988(2)	7/2+	+2.3705(7)	+317.0(20)*
74	W	180	179.946 704(4)	0.12(1)	0+	0	
		182	181.948 204 2(9)	26.50(16)	0+	0	
		183	182.950 223 0(9)	14.31(4)	1/2−	+0.117 784 76(9)	
		184	183.950 931 2(9)	30.64(2)	0+	0	
		186	185.954 364 1(19)	28.43(19)	0+	0	
75	Re	185	184.952 955 0(13)	37.40(2)	5/2+	+3.1871(3)	+218.0(20)*
		187#	186.955 753 1(15)	62.60(2)	5/2+	+3.2197(3)	+207.0(20)*
76	Os	184	183.952 489 1(14)	0.02(1)	0+	0	
		186#	185.953 838 2(15)	1.59(3)	0+	0	
		187	186.955 750 5(15)	1.96(2)	1/2−	+0.064 651 89(6)*	
		188	187.955 838 2(15)	13.24(8)	0+	0	
		189	188.958 147 5(16)	16.15(5)	3/2−	+0.659 933(4)	+98.0(60)**
		190	189.958 447 0(16)	26.26(2)	0+	0	
		192	191.961 480 7(27)	40.78(19)	0+	0	
77	Ir	191	190.960 594 0(18)	37.3(2)	3/2+	+0.1507(6)*	+81.6(9)*
		193	192.962 926 4(18)	62.7(2)	3/2+	+0.1637(6)*	+75.1(9)*
78	Pt	190#	189.959 932(6)	0.014(1)	0+	0	
		192	191.961 038 0(27)	0.782(7)	0+	0	
		194	193.962 680 3(9)	32.967(99)	0+	0	
		195	194.964 791 1(9)	33.832(10)	1/2−	+0.609 52(6)	
		196	195.964 951 5(9)	25.242(41)	0+	0	
		198	197.967 893(3)	7.163(55)	0+	0	
79	Au	197	196.966 568 7(6)	100	3/2+	+0.145 746(9)**	+54.7(16)**
80	Hg	196	195.965 833(3)	0.15(1)	0+	0	
		198	197.966 769 0(4)	9.97(20)	0+	0	
		199	198.968 279 9(4)	16.87(22)	1/2−	+0.505 885 5(9)	
		200	199.968 326 0(4)	23.10(19)	0+	0	
		201	200.970 302 3(6)	13.18(9)	3/2−	−0.560 225 7(14)*	+38.0(40)*
		202	201.970 643 0(6)	29.86(26)	0+	0	
		204	203.973 493 9(4)	6.87(15)	0+	0	
81	Tl	203	202.972 344 2(14)	29.52(1)	1/2+	+1.622 257 87(12)*	
		205	204.974 427 5(14)	70.48(1)	1/2+	+1.638 214 61(12)	
82	Pb	204	203.973 043 6(13)	1.4(1)	0+	0	
		206	205.974 465 3(13)	24.1(1)	0+	0	
		207	206.975 896 9(13)	22.1(1)	1/2−	+0.592 583(9)*	
		208	207.976 652 1(13)	52.4(1)	0+	0	

Z	Symbol	A	Atomic mass, m_a/u	Isotopic composition, $100\,x$	Nuclear spin, I	Magnetic moment, m/μ_N	Quadrupole moment, Q/fm^2
83	Bi	209#	208.980 398 7(16)	100	9/2−	+4.1103(5)*	−51.6(15)*
84	Po	209*	208.982 430 4(20)		1/2−		
85	At	210*	209.987 148(8)		(5)+		
86	Rn	222*	222.017 577 7(25)		0+	0	
87	Fr	223*	223.019 735 9(26)		3/2(−)	+1.17(2)	+117.0(10)
88	Ra	226*	226.025 409 8(25)		0+	0	
89	Ac	227*	227.027 752 1(26)		3/2−	+1.1(1)	+170.0(200)
90	Th	232#	232.038 055 3(21)	100	0+	0	
91	Pa	231*	231.035 884 0(24)	100	3/2−	2.01(2)	
92	U	233*	233.039 635 2(29)		5/2+	0.59(5)	366.3(8)*
		234#	234.040 952 2(20)	0.0054(5)	0+	0	
		235#	235.043 929 9(20)	0.7204(6)	7/2−	−0.38(3)*	493.6(6)*
		238#	238.050 788 2(20)	99.2742(10)	0+	0	
93	Np	237*	237.048 173 4(20)		5/2+	+3.14(4)*	+386.6(6)
94	Pu	244*	244.064 204(5)		0+		
95	Am	243*	243.061 381 1(25)		5/2−	+1.503(14)	+286.0(30)*
96	Cm	247*	247.070 354(5)		9/2−	0.36(7)	
97	Bk	247*	247.070 307(6)				
98	Cf	251*	251.079 587(5)				
99	Es	252*	252.082 980(50)				
		253*	253.084 824 7(28)		7/2+	+4.10(7)	670.0(800)
100	Fm	257*	257.095 105(7)				
101	Md	258*	258.098 431(5)				
102	No	259*	259.101 03(11)				
103	Lr	262*	262.109 63(22)				
104	Rf	261*	261.108 77(3)				
105	Db	262*	262.114 08(20)				
106	Sg	263*	263.118 32(13)				
107	Bh	264*	264.1246(3)				
108	Hs	265*	265.130 09(15)				
109	Mt	268*	268.138 73(34)				
110	Ds	271*	271.146 06(11)				
111	Rg	272*	272.153 62(36)				

7 CONVERSION OF UNITS

Units of the SI are recommended for use throughout science and technology. However, some non-rationalized units are in use, and in a few cases they are likely to remain so for many years. Moreover, the published literature of science makes widespread use of non-SI units. It is thus often necessary to convert the values of physical quantities between SI units and other units. This chapter is concerned with facilitating this process, as well as the conversion of units in general.

Section 7.1, p. 131 gives examples illustrating the use of quantity calculus for converting the numerical values of physical quantities expressed in different units. The table in Section 7.2, p. 135 lists a variety of non-rationalized units used in chemistry, with the conversion factors to the corresponding SI units. Transformation factors for energy and energy-related units (repetency, wavenumber, frequency, temperature and molar energy), and for pressure units, are also presented in tables in the back of this manual.

Many of the difficulties in converting units between different systems are associated either with the electromagnetic units, or with atomic units and their relation to the electromagnetic units. In Sections 7.3 and 7.4, p. 143 and 146 the relations involving electromagnetic and atomic units are developed in greater detail to provide a background for the conversion factors presented in the table in Section 7.2, p. 135.

7.1 THE USE OF QUANTITY CALCULUS

Quantity calculus is a system of algebra in which symbols are consistently used to represent physical quantities and not their numerical values expressed in certain units. Thus we always take the values of physical quantities to be the product of a numerical value and a unit (see Section 1.1, p. 3), and we manipulate the symbols for physical quantities, numerical values, and units by the ordinary rules of algebra (see footnote [1], below). This system is recommended for general use in science and technology. Quantity calculus has particular advantages in facilitating the problems of converting between different units and different systems of units. Another important advantage of quantity calculus is that equations between quantities are independent of the choice of units, and must always satisfy the rule that the dimensions must be the same for each term on either side of the equal sign. These advantages are illustrated in the examples below, where the numerical values are approximate.

Example 1. The wavelength λ of one of the yellow lines of sodium is given by

$$\lambda \approx 5.896 \times 10^{-7} \text{ m}, \quad \text{or} \quad \lambda/\text{m} \approx 5.896 \times 10^{-7}$$

The ångström is defined by the equation (see Section 7.2, "length", p. 135)

$$1 \text{ Å} = \text{Å} := 10^{-10} \text{ m}, \quad \text{or} \quad \text{m}/\text{Å} := 10^{10}$$

Substituting in the first equation gives the value of λ in ångström

$$\lambda/\text{Å} = (\lambda/\text{m}) \, (\text{m}/\text{Å}) \approx (5.896 \times 10^{-7}) \, (10^{10}) = 5896$$

or

$$\lambda \approx 5896 \text{ Å}$$

Example 2.   The vapor pressure of water at 20 °C is recorded to be

$$p(\text{H}_2\text{O}, \, 20 \text{ °C}) \approx 17.5 \text{ Torr}$$

The torr, the bar, and the atmosphere are given by the equations (see Section 7.2, "pressure", p. 138)

1 Torr	$\approx$	133.3 Pa
1 bar	$:=$	10^5 Pa
1 atm	$:=$	101 325 Pa

Thus

$$\begin{aligned} p(\text{H}_2\text{O}, \, 20 \text{ °C}) &\approx 17.5 \times 133.3 \text{ Pa} \approx 2.33 \text{ kPa} = \\ &(2.33 \times 10^3/10^5) \text{ bar} = 23.3 \text{ mbar} = \\ &(2.33 \times 10^3/101\,325) \text{ atm} \approx 2.30 \times 10^{-2} \text{ atm} \end{aligned}$$

Example 3.   Spectroscopic measurements show that for the methylene radical, CH_2, the $\tilde{\text{a}} \, {}^1\text{A}_1$ excited state lies at a repetency (wavenumber) 3156 cm^{-1} above the $\tilde{\text{X}} \, {}^3\text{B}_1$ ground state

$$\tilde{\nu}(\tilde{\text{a}} - \tilde{\text{X}}) = T_0(\tilde{\text{a}}) - T_0(\tilde{\text{X}}) \approx 3156 \text{ cm}^{-1}$$

The excitation energy from the ground triplet state to the excited singlet state is thus

$$\begin{aligned} \Delta E = hc_0\tilde{\nu} &\approx (6.626 \times 10^{-34} \text{ J s}) \, (2.998 \times 10^8 \text{ m s}^{-1}) \, (3156 \text{ cm}^{-1}) \approx \\ &6.269 \times 10^{-22} \text{ J m cm}^{-1} = \\ &6.269 \times 10^{-20} \text{ J} = 6.269 \times 10^{-2} \text{ aJ} \end{aligned}$$

---

[1] A more appropriate name for "quantity calculus" might be "algebra of quantities", because the principles of algebra rather than calculus (in the sense of differential and integral calculus) are involved.

where the values of h and c_0 are taken from the fundamental physical constants in Chapter 5, p. 111 and we have used the relation 1 m = 100 cm, or 1 m 1 cm^{-1} = 100. Since the electronvolt is given by the equation (Section 7.2, "energy", p. 137) 1 eV $\approx 1.6022 \times 10^{-19}$ J, or 1 aJ $\approx$ (1/0.160 22) eV,

$$\Delta E \approx \left(6.269 \times 10^{-2}/0.160\ 22\right)\ \text{eV} \approx 0.3913\ \text{eV}$$

Similarly the hartree is given by $E_h = \hbar^2/m_e a_0{}^2 \approx 4.3597$ aJ, or 1 aJ $\approx (1/4.3597)E_h$ (Section 3.9.1, p. 94), and thus the excitation energy is given in atomic units by

$$\Delta E \approx \left(6.269 \times 10^{-2}/4.3597\right) E_h \approx 1.4379 \times 10^{-2} E_h$$

Finally the molar excitation energy is given by

$$\Delta E_m = L\Delta E \approx (6.022 \times 10^{23}\ \text{mol}^{-1})(6.269 \times 10^{-2}\ \text{aJ}) \approx 37.75\ \text{kJ mol}^{-1}$$

Also, since 1 kcal := 4.184 kJ, or 1 kJ := (1/4.184) kcal,

$$\Delta E_m \approx (37.75/4.184)\ \text{kcal mol}^{-1} \approx 9.023\ \text{kcal mol}^{-1}$$

In the transformation from ΔE to ΔE_m the coefficient L is not a number, but has a dimension different from one. Also in this example the necessary transformation coefficient could have been taken directly from the table in the back of this manual.

Example 4.   The molar conductivity, Λ, of an electrolyte is defined by the equation

$$\Lambda = \kappa/c$$

where κ is the conductivity of the electrolyte solution minus the conductivity of the pure solvent and c is the electrolyte concentration. Conductivities of electrolytes are usually expressed in S cm^{-1} and concentrations in mol dm^{-3}; for example $\kappa(\text{KCl}) \approx 7.39 \times 10^{-5}$ S cm^{-1} for $c(\text{KCl}) \approx 0.000\ 500$ mol dm^{-3}. The molar conductivity can then be calculated as follows

$$\Lambda \approx (7.39 \times 10^{-5}\ \text{S cm}^{-1})/(0.000\ 500\ \text{mol dm}^{-3}) \approx$$
$$0.1478\ \text{S mol}^{-1}\ \text{cm}^{-1}\ \text{dm}^3 = 147.8\ \text{S mol}^{-1}\ \text{cm}^2$$

since 1 dm^3 = 1000 cm^3. The above relation has previously often been, and sometimes still is, written in the not-recommended form

$$\Lambda = 1000\ \kappa/c$$

However, in this form the symbols *do not* represent physical quantities, but the *numerical values* of physical quantities expressed in certain units. Specifically, the last equation is true only if Λ is the numerical value of the molar conductivity in S mol^{-1} cm^2, κ is the numerical value of the conductivity in S cm^{-1}, and c is the numerical value of the concentration in mol dm^{-3}. This form does not follow the rules of quantity calculus, and should be avoided. The equation $\Lambda = \kappa/c$, in which the symbols represent physical quantities, is true in any units. If it is desired to write the relation between numerical values it should be written in the form

$$\Lambda/(\text{S mol}^{-1}\ \text{cm}^2) = \frac{1000\ \kappa/(\text{S cm}^{-1})}{c/(\text{mol dm}^{-3})}$$

Example 5.   A solution of 0.125 mol of solute B in $m_S \approx 953$ g of solvent S has a molality b_B given by (see also footnote [2], below)

$$b_B = n_B/m_S \approx (0.125/953) \text{ mol g}^{-1} \approx 0.131 \text{ mol kg}^{-1}$$

The amount-of-substance fraction of solute is approximately given by

$$x_B = n_B/(n_S + n_B) \approx n_B/n_S = b_B M_S$$

where it is assumed that $n_B \ll n_S$.

If the solvent is water with molar mass 18.015 g mol^{-1}, then

$$x_B \approx (0.131 \text{ mol kg}^{-1})(18.015 \text{ g mol}^{-1}) \approx 2.36 \text{ g/kg} = 0.002\ 36$$

These equations are sometimes quoted in the deprecated form $b_B = 1000\,n_B/M_S$, and $x_B \approx b_B M_S/1000$. However, this is *not* a correct use of quantity calculus because in this form the symbols denote the *numerical values* of the physical quantities in particular units; specifically it is assumed that b_B, m_S and M_S denote numerical values in mol kg^{-1}, g, and g mol^{-1} respectively. A correct way of writing the second equation would, for example, be

$$x_B = (b_B/\text{mol kg}^{-1})(M_S/\text{g mol}^{-1})/1000$$

Example 6.   For paramagnetic materials the magnetic susceptibility may be measured experimentally and used to give information on the molecular magnetic dipole moment, and hence on the electronic structure of the molecules in the material. The paramagnetic contribution to the molar magnetic susceptibility of a material, χ_m, is related to the molecular magnetic dipole moment m by the Curie relation

$$\chi_m = \chi V_m = \mu_0 N_A m^2/3kT$$

In the older non-rationalized esu, emu, and Gaussian systems (see Section 7.3, p. 143), this equation becomes

$$\chi_m^{(ir)} = \chi^{(ir)} V_m = N_A m^2/3kT$$

Solving for m, and expressing the result in terms of the Bohr magneton μ_B, in the ISQ(SI)

$$m/\mu_B = (3k/\mu_0 N_A)^{1/2} \mu_B^{-1} (\chi_m T)^{1/2}$$

and in the non-rationalized emu and Gaussian systems

$$m/\mu_B = (3k/N_A)^{1/2} \mu_B^{-1} \left(\chi_m^{(ir)} T\right)^{1/2}$$

Finally, using the values of the fundamental physical constants μ_B, k, μ_0, and N_A given in Chapter 5, p. 111, we obtain

$$m/\mu_B \approx 0.7977 \left[\chi_m/ (\text{cm}^3 \text{ mol}^{-1})\right]^{1/2} [T/K]^{1/2} \approx 2.828 \left[\chi_m^{(ir)}/ (\text{cm}^3 \text{ mol}^{-1})\right]^{1/2} [T/K]^{1/2}$$

These expressions are convenient for practical calculations. The final result has frequently been expressed in the not recommended form

$$m/\mu_B \approx 2.828 \left(\chi_m T\right)^{1/2}$$

where it is assumed, contrary to the conventions of quantity calculus, that χ_m and T denote the *numerical values* of the molar susceptibility and the temperature in the units cm^3 mol^{-1} and K, respectively, and where it is also assumed (but rarely stated) that the susceptibility is defined using the electromagnetic equations defined within non-rationalized systems (see Section 7.3, p. 143).

---

[2] We use b_B because of the possible confusion of notation when using m_B to denote molality, and m_S to denote the mass of S. However, the symbol m_B is frequently used to denote molality (see Section 2.10, note 14, p. 48).

Example 7.   The esu or Gaussian unit of the electric charge is $\sqrt{\mathrm{erg\,cm}}$. Alternatively, the franklin (Fr) is used to represent $\sqrt{\mathrm{erg\,cm}}$ (see Section 7.3, p. 143). The definition of the franklin states that two particles, each having an electric charge of one franklin and separated by a distance of one centimetre, repel each other with a force of one dyn. In order to obtain a conversion factor to the SI unit coulomb, one needs to ask what charges produce the same force in SI units at the same distance. One has thus

$$1\,\mathrm{dyn} = 1\,\mathrm{cm\,g\,s^{-2}} = 10^{-5}\,\mathrm{N} = \frac{1}{4\pi\varepsilon_0}\frac{Q^2}{(10^{-2}\,\mathrm{m})^2}$$

It follows (see Section 7.3, p. 143) that

$$Q^2 = 4\pi\varepsilon_0 \times 10^{-9}\,\mathrm{N\,m^2} = \mu_0\varepsilon_0 \times 10^{-2}\,\mathrm{A^2\,m^2} = c_0^{-2} \times 10^{-2}\,\mathrm{A^2\,m^2}$$

and thus

$$Q = \frac{1}{10\,c_0}\,\mathrm{A\,m} = \frac{10}{\zeta}\,\mathrm{C}$$

where ζ is the exact number $\zeta = c_0/(\mathrm{cm\,s^{-1}}) = 29\,979\,245\,800$ (see Sections 7.2 and 7.3, p. 135 and 143). Thus 1 Fr $= 1/(2.997\,924\,58 \times 10^9)$ C.

Example 8.   The esu or Gaussian unit of the electric field strength is $\sqrt{\mathrm{erg\,cm}}\,\mathrm{cm^{-2}}$ or, alternatively, Fr cm^{-2}. Simple insertion of the above mentioned relation between the franklin and the coulomb would yield 1 Fr cm$^{-2} = (10^5/\zeta)$ C m^{-2}. However, C m^{-2} is not the unit of the electric field strength in the SI. In order to obtain the conversion factor for the electric field strength in the SI, one has to set

$$1\,\frac{\mathrm{Fr}}{\mathrm{cm^2}} = \frac{1}{4\pi\varepsilon_0} \times \frac{10^5}{\zeta}\,\frac{\mathrm{C}}{\mathrm{m^2}}$$

From Example 7 one has $4\pi\varepsilon_0 \times 10^{-9}\,\mathrm{N\,m^2} = (10/\zeta)^2\,\mathrm{C^2}$, thus

$$\frac{10^5\,\mathrm{C}}{4\pi\varepsilon_0\zeta\mathrm{m^2}} = \zeta \times 10^{-6}\frac{\mathrm{N}}{\mathrm{C}}$$

One concludes that 1 Fr cm$^{-2} = \zeta \times 10^{-6}$ V m^{-1}.

Example 9.   The esu or Gaussian unit of the polarizability is cm^3: in a system of charges with polarizability 1 cm^3, a field of 1 Fr/cm^2 produces a dipole moment of 1 Fr cm (see Section 7.3, p. 143). The corresponding value of this polarizability in the SI is obtained as follows:

$$1\,\mathrm{cm^3} = \frac{p}{E} = \frac{(10/\zeta)\,\mathrm{C}\,10^{-2}\,\mathrm{m}}{\zeta\,10^{-6}\,\mathrm{V\,m^{-1}}} = \zeta^{-2}\,10^5\,\mathrm{m^2\,C\,V^{-1}}$$

One has thus 1 cm$^3 = (10^5/\zeta^2)\,\mathrm{m^2\,C^2\,J^{-1}}$, from the definition of the volt. This equation seems strange, at a first sight. It does not give a general relation between a volume of 1 cm^3 and other units in the SI. Rather it states that a polarizability that has the value 1 cm^3 in the esu or Gaussian system has the value $(10^5/\zeta^2)\,\mathrm{m^2\,C^2\,J^{-1}}$ in the SI.

134

7.2 CONVERSION TABLES FOR UNITS

The table below gives conversion factors from a variety of units to the corresponding SI unit. Examples of the use of this table have already been given in the preceding section. For each physical quantity the name is given, followed by the recommended symbol(s). Then the SI unit is given, followed by the esu, emu, Gaussian unit, atomic unit (au), and other units in common use, with their conversion factors to SI and other units. The constant ζ which occurs in some of the electromagnetic conversion factors is the exact number $29\ 979\ 245\ 800 = c_0/(\text{cm s}^{-1})$.

The inclusion of non-SI units in this table should not be taken to imply that their use is to be encouraged. With some exceptions, SI units are always to be preferred to non-SI units. However, since many of the units below are to be found in the scientific literature, it is convenient to tabulate their relation to the SI.

For convenience units in the esu, emu and Gaussian systems are quoted in terms of the four dimensions *length, mass, time,* and *electric charge,* by including the franklin (Fr) as the electrostatic unit of charge and the biot (Bi) as the electromagnetic unit of current. This gives each physical quantity the same dimensions in all systems, so that all conversion factors are numbers. The factors $4\pi\varepsilon_0 = k_{es}^{-1}$ and the Fr may be eliminated by writing $\text{Fr} = \text{erg}^{1/2}\,\text{cm}^{1/2} = \text{cm}^{3/2}\,\text{g}^{1/2}\,\text{s}^{-1}$, and $k_{es} = 1\,\text{Fr}^{-2}\,\text{erg cm} = 1$, to recover esu expressions in terms of three base units (see Section 7.3, p. 143). The symbol Fr should be regarded as a symbol for the electrostatic unit of charge. Similarly, the factor $\mu_0/4\pi = k_{em}$ and the Bi may be eliminated by using $\text{Bi} = \text{dyn}^{1/2} = \text{cm}^{1/2}\,\text{g}^{1/2}\,\text{s}^{-1}$, and $k_{em} = \text{dyn Bi}^{-2} = 1$, to recover emu expressions in terms of the three base units.

The table must be read in the following way, for example for the fermi, as a unit of length: $\text{f} = 10^{-15}$ m means that the symbol f of a length 13.1 f may be replaced by 10^{-15} m, saying that the length has the value 13.1×10^{-15} m. A more difficult example involves for instance the polarizability α. When α is 17.8 cm^3 in the Gaussian system, it is $(17.8 \times 10^5/\zeta^2)\,\text{m}^2\,\text{C}^2\,\text{J}^{-1} \approx 3.53 \times 10^{-9}\,\text{m}^2\,\text{C}^2\,\text{J}^{-1}$ in the SI. That is, the symbol cm^3 may be replaced by the expression $(10^5/\zeta^2)\,\text{m}^2\,\text{C}^2\,\text{J}^{-1}$ to obtain the value of the polarizability in the SI (see also the Example 9 in Section 7.1, p. 134). Conversion factors are either given exactly (when the $=$ sign is used; if it is a definition, the sign $:=$ is used), or they are given to the approximation that the corresponding physical constants are known (when the sign $\approx$ is used). In the latter case the magnitude of the uncertainty is always less than 5 in the last digit quoted, if the uncertainty is not explicitly indicated.

Name	Symbol	Expressed in SI units	Notes
length, l			
metre (SI unit)	m		
centimetre (CGS unit)	cm	$= 10^{-2}$ m	
bohr (au)	a_0	$= 4\pi\varepsilon_0\hbar^2/m_e e^2 \approx 5.291\ 772\ 085\ 9(36)\times10^{-11}$ m	
ångström	Å	$= 10^{-10}$ m	
micron	μ	$= 1\ \mu\text{m} = 10^{-6}$ m	
millimicron	mμ	$= 1\ \text{nm} = 10^{-9}$ m	
x unit	X	$\approx 1.002\times10^{-13}$ m	
fermi	f	$= 1\ \text{fm} = 10^{-15}$ m	
inch	in	$= 2.54\times10^{-2}$ m	
foot	ft	$= 12\ \text{in} = 0.3048$ m	
yard	yd	$= 3\ \text{ft} = 0.9144$ m	
mile	mi	$= 1760\ \text{yd} = 1609.344$ m	
nautical mile		$= 1852$ m	

Name	Symbol	Expressed in SI units	Notes
length, l(continued)			
astronomical unit	ua	$\approx 1.495\ 978\ 706\ 91(6)\times10^{11}$ m	
parsec	pc	$\approx 3.085\ 678\times10^{16}$ m	
light year	l.y.	$\approx 9.460\ 730\times10^{15}$ m	1
light second		$-$ 299 792 458 m	
area, A			
square metre (SI unit)	m^2		
barn	b	$= 10^{-28}$ m^2	
acre		≈ 4046.856 m^2	
are	a	$= 100$ m^2	
hectare	ha	$= 10^4$ m^2	
volume, V			
cubic metre (SI unit)	m^3		
litre	l, L	$= 1$ dm$^3 = 10^{-3}$ m^3	2
lambda	λ	$= 10^{-6}$ dm^3 $[= 1\ \mu l]$	
barrel (US)		$= 158.9873$ dm^3	
gallon (US)	gal (US)	$= 3.785\ 412$ dm^3	
gallon (UK)	gal (UK)	$- 4.546\ 092$ dm^3	
plane angle, α			
radian (SI unit)	rad		
degree	°, deg	$= (\pi/180)$ rad $\approx (1/57.295\ 78)$ rad	
minute	′	$= (\pi/10\ 800)$ rad $[= (1/60)°]$	
second	″	$= (\pi/648\ 000)$ rad $[= (1/3600)°]$	
gon	gon	$= (\pi/200)$ rad $\approx (1/63.661\ 98)$ rad	
mass, m			
kilogram (SI unit)	kg		
gram (CGS unit)	g	$= 10^{-3}$ kg	
electron mass (au)	m_e	$\approx 9.109\ 382\ 15(45)\times10^{-31}$ kg	
dalton,	Da, u	$:= m_a(^{12}C)/12 \approx 1.660\ 538\ 782(83)\times10^{-27}$ kg	
unified atomic mass unit			
gamma	γ	$- 1\ \mu g$	
tonne, (metric tonne)	t	$= 1$ Mg $= 10^3$ kg	
pound (avoirdupois)	lb	$= 0.453\ 592\ 37$ kg	
pound, (metric pound)		$= 0.5$ kg	
ounce (avoirdupois)	oz	$= 28.349\ 52$ g	
ounce (troy)	oz (troy)	$= 31.103\ 476\ 8$ g	
grain	gr	$- 64.798\ 91$ mg	

(1) This value refers to the Julian year (1 a $=$ 365.25 d, 1 d $=$ 86 400 s) [152]. l.y. is not a symbol according to the syntax rules of Section 1.3, p. 5, but is an often used abbreviation.
(2) ISO and IEC only use the lower case l for the litre.

Name	Symbol	Expressed in SI units	Notes
time t			
second (SI unit, CGS unit)	s		
au of time	$\hbar/E_h$	$\approx 2.418\ 884\ 326\ 505(16){\times}10^{-17}$ s	
minute	min	$= 60$ s	
hour	h	$= 3600$ s	
day	d	$= 24$ h $= 86\ 400$ s	3
year	a	$\approx 31\ 556\ 952$ s	4
svedberg	Sv	$= 10^{-13}$ s	
acceleration, a			
SI unit	m s^{-2}		
standard acceleration of free fall	g_n	$= 9.806\ 65$ m s^{-2}	
gal (CGS unit)	Gal	$= 10^{-2}$ m s^{-2}	
force, F			
newton (SI unit)	N	$= 1$ kg m s^{-2}	5
dyne (CGS unit)	dyn	$= 1$ g cm s$^{-2} = 10^{-5}$ N	
au of force	E_h/a_0	$\approx 8.238\ 722\ 06(41){\times}10^{-8}$ N	
kilogram-force, kilopond	kgf $=$ kp	$= 9.806\ 65$ N	
energy, E, U			
joule (SI unit)	J	$= 1$ kg m^2 s^{-2}	
erg (CGS unit)	erg	$= 1$ g cm^2 s$^{-2} = 10^{-7}$ J	
hartree (au)	E_h	$= \hbar^2/m_e a_0^2 \approx 4.359\ 743\ 94(22){\times}10^{-18}$ J	
rydberg	Ry	$= E_h/2 \approx 2.179\ 871\ 97(11){\times}10^{-18}$ J	
electronvolt	eV	$= e{\cdot}1$ V $\approx 1.602\ 176\ 487(40){\times}10^{-19}$ J	
calorie, thermochemical	cal$_{th}$	$= 4.184$ J	
calorie, international	cal$_{IT}$	$= 4.1868$ J	
15 °C calorie	cal$_{15}$	≈ 4.1855 J	
litre atmosphere	l atm	$= 101.325$ J	
British thermal unit	Btu	$= 1055.06$ J	

(3) Note that the astronomical day is not exactly defined in terms of the second since so-called leap-seconds are added or subtracted from the day semiannually in order to keep the annual average occurrence of midnight at 24:00:00 on the clock. However, in reference [3] hour and day are defined in terms of the second as given.

(4) The year is not commensurable with the day and not a constant. Prior to 1967, when the atomic standard was introduced, the tropical year 1900 served as the basis for the definition of the second. For the epoch 1900.0 it amounted to 365.242 198 79 d $\approx$ 31 556 925.98 s and it decreases by 0.530 s per century. The calendar years are exactly defined in terms of the day: The **Julian** year is equal to 365.25 d; the **Gregorian** year is equal to 365.2425 d; the **Mayan** year is equal to 365.2420 d. The definition in the table corresponds to the Gregorian year. This is an average based on a year of length 365 d, with leap years of 366 d; leap years are taken *either* when the year is divisible by 4 but is not divisible by 100, or when the year is divisible by 400. Whether the year 3200 should be a leap year is still open, but this does not have to be resolved until sometime in the middle of the 32nd century. For conversion one may use in general approximately 1 a $\approx 3.1557{\times}10^7$ s. If more accurate statements are needed, the precise definition of the year used should be stated.

Name	Symbol	Expressed in SI units	Notes
pressure, p			
pascal (SI unit)	Pa	$= 1$ N m$^{-2} = 1$ kg m^{-1} s^{-2}	
standard atmosphere	atm	$= 101\ 325$ Pa	
bar	bar	$= 10^5$ Pa	
torr	Torr	$= (101\ 325/760)$ Pa ≈ 133.322 Pa	
conventional millimetre of mercury	mmHg	$= 13.5951 \times 980.665 \times 10^{-2}$ Pa $\approx$ 133.322 Pa	
pounds per square inch	psi	$\approx 6.894\ 758 \times 10^3$ Pa	
power, P			
watt (SI unit)	W	$= 1$ kg m^2 s^{-3}	
Imperial horse power	hp	≈ 745.7 W	
metric horse power	hk	$= 735.498\ 75$ W	
action, angular momentum, L, J			
SI unit	J s	$= 1$ kg m^2 s^{-1}	
CGS unit	erg s	$= 10^{-7}$ J s	
au of action	$\hbar$	$:= h/2\pi \approx 1.054\ 571\ 628(53) \times 10^{-34}$ J s	
dynamic viscosity, η			
SI unit	Pa s	$= 1$ kg m^{-1} s^{-1}	
poise (CGS unit)	P	$= 10^{-1}$ Pa s	
centipoise	cP	$= 1$ mPa s	
kinematic viscosity, ν			
SI unit	m^2 s^{-1}		
stokes (CGS unit)	St	$= 10^{-4}$ m^2 s^{-1}	
thermodynamic temperature, T			
kelvin (SI unit)	K		
degree Rankine	°R	$= (5/9)$ K	6
Celsius temperature, t			
degree Celsius (SI unit)	°C	$= 1$ K	6
entropy, S			
heat capacity, C			
SI unit	J K^{-1}		
clausius	Cl	$= 4.184$ J K^{-1} $[= 1$ cal$_{th}$ K$^{-1}]$	

(Notes continued)

(5) 1 N is approximately the force exerted by the Earth upon an apple.

(6) $T/°\mathrm{R} = (9/5)T/\mathrm{K}$. Also, Celsius temperature t is related to thermodynamic temperature T by the equation

$$t/°\mathrm{C} = T/\mathrm{K} - 273.15$$

Similarly Fahrenheit temperature t_F is related to Celsius temperature t by the equation

$$t_F/°\mathrm{F} = (9/5)(t/°\mathrm{C}) + 32$$

Name	Symbol	Expressed in SI units	Notes
molar entropy, S_m			
molar heat capacity, C_m			
SI unit	J K^{-1} mol^{-1}	$= 1$ kg m^2 s^{-2} K^{-1} mol^{-1}	
entropy unit		$= 4.184$ J K^{-1} mol^{-1} $=$ [1 cal$_{th}$ K^{-1} mol^{-1}]	
molar volume, V_m			
SI unit	m^3 mol^{-1}		
amagat		$:= V_m$ of real gas at 1 atm and 273.15 K $\approx 22.4 \times 10^{-3}$ m^3 mol^{-1}	
amount density, $1/V_m$			
SI unit	mol m^{-3}		
reciprocal amagat		$:= 1/V_m$ of real gas at 1 atm and 273.15 K ≈ 44.6 mol m^{-3}	7
activity, A			
becquerel (SI unit)	Bq	$= 1$ s^{-1}	
curie	Ci	$= 3.7 \times 10^{10}$ Bq	
absorbed dose of radiation, D			8
gray (SI unit)	Gy	$= 1$ J kg$^{-1} = 1$ m^2 s^{-2}	
rad	rad	$= 0.01$ Gy	
dose equivalent, H			
sievert (SI unit)	Sv	$= 1$ J kg$^{-1} = 1$ m^2 s^{-2}	
rem	rem	$= 0.01$ Sv	9
electric current, I			
ampere (SI unit)	A		
esu, Gaussian	Fr s^{-1}	$:= (10/\zeta)$A $\approx 3.335\ 64 \times 10^{-10}$ A	10
biot (emu)	Bi	$= 10$ A	
au	$eE_h/\hbar$	$\approx 6.623\ 617\ 63(17) \times 10^{-3}$ A	
electric charge, Q			
coulomb (SI unit)	C	$= 1$ A s	
franklin (esu, Gaussian)	Fr	$:= (10/\zeta)$C $\approx 3.335\ 64 \times 10^{-10}$ C	10
emu (abcoulomb)	Bi s	$= 10$ C	
proton charge (au)	e	$\approx 1.602\ 176\ 487(40) \times 10^{-19}$ C $\approx 4.803\ 21 \times 10^{-10}$ Fr	

(7) The name "amagat" is unfortunately often used as a unit for both molar volume and amount density. Its value is slightly different for different gases, reflecting the deviation from ideal behaviour for the gas being considered.

(8) The unit röntgen, R, is employed to express exposure to X or γ radiation. 1 R $= 2.58 \times 10^{-4}$ C kg^{-1}.

(9) rem stands for röntgen equivalent man.

(10) ζ is the exact number $\zeta = c_0/(\text{cm s}^{-1}) = 29\ 979\ 245\ 800$.

Name	Symbol	Expressed in SI units	Notes
charge density, ρ			
SI unit	C m^{-3}	$= 1$ A s m^{-3}	
esu, Gaussian	Fr cm^{-3}	$:= (10^7/\zeta)$ C m$^{-3} \approx$	10
		$3.335\ 640\ 952{\times}10^{-4}$ C m^{-3}	
au	$ea_0{}^{-3}$	$\approx 1.081\ 202\ 300(27){\times}10^{12}$ C m^{-3}	
electric potential, V, ϕ			
electric tension, U			
volt (SI unit)	V	$= 1$ J C$^{-1} = 1$ kg m^2 s^{-3} A^{-1}	
esu, Gaussian	Fr cm^{-1}	$= \zeta{\times}10^{-8}$ V $= 299.792\ 458$ V	10
"cm^{-1}"	e cm$^{-1}/4\pi\varepsilon_0$	$\approx 1.439\ 964\ 41(36){\times}10^{-7}$ V	11
au	$e/4\pi\varepsilon_0 a_0$	$= E_h/e \approx 27.211\ 383\ 86(68)$ V	
mean international volt		$= 1.000\ 34$ V	
US international volt		$= 1.000\ 330$ V	
electric resistance, R			
ohm (SI unit)	Ω	$= 1$ V A$^{-1} = 1$ m^2 kg s^{-3} A^{-2}	
mean international ohm		$= 1.000\ 49$ Ω	
US international ohm		$= 1.000\ 495$ Ω	
Gaussian	s cm^{-1}	$= \zeta^2 \times 10^{-9}$ $\Omega \approx 8.987\ 551\ 787{\times}10^{11}$ Ω	10
conductivity, κ, σ			
SI	S m^{-1}	$= 1$ kg^{-1} m^{-3} s^3 A^2	
Gaussian	s^{-1}	$= (10^{11}/\zeta^2)$ S m$^{-1} \approx$	10
		$1.112\ 650\ 056{\times}10^{-10}$ S m^{-1}	
capacitance, C			
farad (SI unit)	F	$= 1$ kg^{-1} m^{-2} s^4 A^2	
Gaussian	cm	$:= (10^9/\zeta^2)$ F $\approx 1.112\ 650\ 056{\times}10^{-12}$ F	10
electric field strength, E			
SI unit	V m^{-1}	$= 1$ J C^{-1} m$^{-1} = 1$ kg m s^{-3} A^{-1}	
esu, Gaussian	Fr^{-1} cm^{-2}	$:= \zeta \times 10^{-6}$ V m$^{-1} =$	10
		$2.997\ 924\ 58{\times}10^4$ V m^{-1}	
"cm^{-2}"	e cm$^{-2}/4\pi\varepsilon_0$	$\approx 1.439\ 964\ 41(36){\times}10^{-5}$ V m^{-1}	11
au	$e/4\pi\varepsilon_0 a_0{}^2$	$= E_h/ea_0 \approx$	
		$5.142\ 206\ 32(13){\times}10^{11}$ V m^{-1}	

(11) The units in quotation marks for electric potential through polarizability may be found in the literature, although they are strictly incorrect; the entry suggested in the column *Symbol* defines the unit in terms of physical quantities and other units, so that, for a conversion into the SI, the physical quantities only need to be replaced by their values in the SI and the units need to be interpreted as units in the SI.

Name	Symbol	Expressed in SI units	Notes
electric field gradient, $E'_{\alpha\beta}$, $q_{\alpha\beta}$			
SI unit	$V\ m^{-2}$	$= 1\ J\ C^{-1}\ m^{-2} = 1\ kg\ s^{-3}\ A^{-1}$	
esu, Gaussian	$Fr^{-1}\ cm^{-3}$	$:= \zeta \times 10^{-4}\ V\ m^{-2} =$ $2.997\ 924\ 58 \times 10^{6}\ V\ m^{-2}$	10
"cm^{-3}"	$e\ cm^{-3}/4\pi\varepsilon_0$	$\approx 1.439\ 964\ 41(36) \times 10^{-3}\ V\ m^{-2}$	11
au	$e/4\pi\varepsilon_0 a_0^{\ 3}$	$= E_h/ea_0^{\ 2} \approx$ $9.717\ 361\ 66(24) \times 10^{21}\ V\ m^{-2}$	
electric dipole moment, p, μ			
SI unit	$C\ m$	$= 1\ A\ s\ m$	
esu, Gaussian	$Fr\ cm$	$:= (10^{-1}/\zeta)\ C\ m \approx$ $3.335\ 640\ 952 \times 10^{-12}\ C\ m$	10
debye	D	$= 10^{-18}\ Fr\ cm \approx$ $3.335\ 640\ 952 \times 10^{-30}\ C\ m$	
"cm", dipole length	$e\ cm$	$\approx 1.602\ 176\ 487(40) \times 10^{-21}\ C\ m$	11
au	ea_0	$\approx 8.478\ 352\ 81(21) \times 10^{-30}\ C\ m$	
electric quadrupole moment, $Q_{\alpha\beta}$, $\Theta_{\alpha\beta}$, eQ			
SI unit	$C\ m^2$	$= 1\ A\ s\ m^2$	
esu, Gaussian	$Fr\ cm^2$	$:= (10^{-3}/\zeta)\ C\ m^2 \approx$ $3.335\ 640\ 952 \times 10^{-14}\ C\ m^2$	10
"cm^2", quadrupole area	$e\ cm^2$	$\approx 1.602\ 176\ 487(40) \times 10^{-23}\ C\ m^2$	11
au	$ea_0^{\ 2}$	$\approx 4.486\ 551\ 07(11) \times 10^{-40}\ C\ m^2$	
polarizability, α			
SI unit	$J^{-1}\ C^2\ m^2$	$= 1\ F\ m^2 = 1\ kg^{-1}\ s^4\ A^2$	
esu, Gaussian	$Fr^2\ cm^3$	$:= (10^5/\zeta^2)\ J^{-1}\ C^2\ m^2 \approx$ $1.112\ 650\ 056 \times 10^{-16}\ J^{-1}\ C^2\ m^2$	10
"cm^3", polarizability volume	$4\pi\varepsilon_0\ cm^3$	$\approx 1.112\ 650\ 056 \times 10^{-16}\ J^{-1}\ C^2\ m^2$	11
"$\mathring{A}^3$"	$4\pi\varepsilon_0\ \mathring{A}^3$	$\approx 1.112\ 650\ 056 \times 10^{-40}\ J^{-1}\ C^2\ m^2$	11
au	$4\pi\varepsilon_0\ a_0^{\ 3}$	$= e^2 a_0^{\ 2}/E_h \approx$ $1.648\ 777\ 253\ 6(34) \times 10^{-41}\ J^{-1}\ C^2\ m^2$	
electric displacement, D *(volume) polarization, P*			12
SI unit	$C\ m^{-2}$	$= 1\ A\ s\ m^{-2}$	
esu, Gaussian	$Fr\ cm^{-2}$	$:= (10^5/\zeta)\ C\ m^{-2} \approx$ $3.335\ 640\ 952 \times 10^{-6}\ C\ m^{-2}$	10
magnetic flux density, B *(magnetic field)*			
tesla (SI unit)	T	$= 1\ J\ A^{-1}\ m^{-2} = 1\ V\ s\ m^{-2} = 1\ Wb\ m^{-2}$	
gauss (emu, Gaussian)	G	$= 10^{-4}\ T$	
au	$\hbar/ea_0^{\ 2}$	$\approx 2.350\ 517\ 382(59) \times 10^5\ T$	

(12) The use of the esu or Gaussian unit for electric displacement usually implies that the non-rationalized displacement is being quoted, $D^{(ir)} = 4\pi D$ (see Section 7.3, p. 143).

Name	Symbol	Expressed in SI units	Notes
magnetic flux, Φ			
weber (SI unit)	Wb	$= 1 \text{ J A}^{-1} = 1 \text{ V s} = 1 \text{ kg m}^2 \text{ s}^{-2} \text{ A}^{-1}$	
maxwell (emu, Gaussian)	Mx	$= 10^{-8} \text{ Wb } [= 1 \text{ G cm}^2]$	
magnetic field strength, H			
SI unit	A m^{-1}		
oersted (emu, Gaussian)	Oe	$= (10^3/4\pi) \text{ A m}^{-1}$	13
(volume) magnetization, M			
SI unit	A m^{-1}		
gauss (emu, Gaussian)	G	$= 10^3 \text{ A m}^{-1}$	13
magnetic dipole moment, m, μ			
SI unit	J T^{-1}	$= 1 \text{ A m}^2$	
emu, Gaussian	erg G^{-1}	$= 10 \text{ A cm}^2 = 10^{-3} \text{ J T}^{-1}$	
Bohr magneton	μ_{B}	$:= e\hbar/2m_{\text{e}} \approx$ $9.274\ 009\ 15(23) \times 10^{-24} \text{ J T}^{-1}$	14
au	$e\hbar/m_{\text{e}}$	$:= 2\mu_{\text{B}} \approx 1.854\ 801\ 830(46) \times 10^{-23} \text{ J T}^{-1}$	
nuclear magneton	μ_{N}	$:= (m_{\text{e}}/m_{\text{p}})\mu_{\text{B}} \approx$ $5.050\ 783\ 24(13) \times 10^{-27} \text{ J T}^{-1}$	
magnetizability, ξ			
SI unit	J T^{-2}	$= 1 \text{ A}^2 \text{ s}^2 \text{ m}^2 \text{ kg}^{-1}$	
Gaussian	erg G^{-2}	$= 10 \text{ J T}^{-2}$	
au	$e^2 a_0{}^2/m_{\text{e}}$	$\approx 7.891\ 036\ 433(27) \times 10^{-29} \text{ J T}^{-2}$	
magnetic susceptibility, χ, κ			
SI unit	1		
emu, Gaussian	1		15
molar magnetic susceptibility, χ_{m}			
SI unit	$\text{m}^3 \text{ mol}^{-1}$		
emu, Gaussian	$\text{cm}^3 \text{ mol}^{-1}$	$= 10^{-6} \text{ m}^3 \text{ mol}^{-1}$	16
inductance, self-inductance, L			
SI unit	H	$= 1 \text{ V s A}^{-1} = 1 \text{ kg m}^2 \text{ s}^{-2} \text{ A}^{-2}$	
Gaussian	$\text{s}^2 \text{ cm}^{-1}$	$= \zeta^2 \times 10^{-9} \text{ H} \approx 8.987\ 551\ 787 \times 10^{11} \text{ H}$	10
emu	cm	$= 10^{-9} \text{ H}$	

(13) In practice the oersted, Oe, is only used as a unit for $H^{(\text{ir})} = 4\pi H$, thus when $H^{(\text{ir})} = 1$ Oe, $H = (10^3/4\pi)$ A m^{-1} (see Section 7.3, p. 143). In the Gaussian or emu system, gauss and oersted are equivalent units.

(14) The Bohr magneton μ_{B} is sometimes denoted BM (or B.M.), but this is not recommended.

(15) In practice susceptibilities quoted in the context of emu or Gaussian units are always values for $\chi^{(\text{ir})} = \chi/4\pi$; thus when $\chi^{(\text{ir})} = 10^{-6}$, $\chi = 4\pi \times 10^{-6}$ (see Section 7.3, p. 143).

(16) In practice the units cm^3 mol^{-1} usually imply that the non-rationalized molar susceptibility is being quoted $\chi_{\text{m}}^{(\text{ir})} = \chi_{\text{m}}/4\pi$; thus, for example if $\chi_{\text{m}}^{(\text{ir})} = -15 \times 10^{-6}$ cm^3 mol^{-1}, which is often written as "-15 cgs ppm", then $\chi_{\text{m}} = -1.88 \times 10^{-10}$ m^3 mol^{-1} (see Section 7.3, p. 143).

7.3 THE ESU, EMU, GAUSSIAN, AND ATOMIC UNIT SYSTEMS IN RELATION TO THE SI

The ISQ (see Section 1.2, p. 4) equations of electromagnetic theory are usually used with physical quantities in SI units, in particular the four base units metre (m), kilogram (kg), second (s), and ampere (A) for length, mass, time, and electric current, respectively. The basic equations for the electrostatic force $\boldsymbol{F}_{\text{es}}$ between particles with charges Q_1 and Q_2 in vacuum, and for the infinitesimal electromagnetic force $\text{d}^2\boldsymbol{F}_{\text{em}}$ between conductor elements of length $\text{d}\boldsymbol{l}_1$ and $\text{d}\boldsymbol{l}_2$ and corresponding currents I_1 and I_2 in vacuum, are

$$\boldsymbol{F}_{\text{es}} \;=\; Q_1Q_2\boldsymbol{r}/4\pi\varepsilon_0 r^3$$
$$\text{d}^2\boldsymbol{F}_{\text{em}} \;=\; (\mu_0/4\pi)\,I_1\,\text{d}\boldsymbol{l}_1 \times (I_2\,\text{d}\boldsymbol{l}_2 \times \boldsymbol{r})/r^3$$

where particles and conductor elements are separated by the vector $\boldsymbol{r}$ (with $|\boldsymbol{r}| = r$). The physical quantity ε_0, the electric constant (formerly called permittivity of vacuum), is defined in the SI system to have the value

$$\varepsilon_0 \;=\; \left(10^7/4\pi c_0{}^2\right)\ \text{kg}^{-1}\,\text{m}^{-1}\,\text{C}^2 \approx 8.854\,187\,817 \times 10^{-12}\ \text{C}^2\,\text{m}^{-1}\,\text{J}^{-1}$$

Similarly, the physical quantity μ_0, the magnetic constant (formerly called the permeability of vacuum), has the value

$$\mu_0 \;:=\; 4\pi \times 10^{-7}\ \text{N A}^{-2} \approx 1.256\,637\,061\,4 \times 10^{-6}\ \text{N A}^{-2}$$

in the SI system. In different unit systems ε_0 and μ_0 may have different numerical values and units. In this book we use ε_0 and μ_0 as shorthands for the SI values as given above and prefer to use more general symbols k_{es} and k_{em} to describe these quantities in other systems of units and equations as discussed below. The SI value of μ_0 results from the definition of the ampere (Section 3.3, p. 87). The value of ε_0 then results from the Maxwell relation

$$\varepsilon_0\mu_0 = 1/c_0{}^2$$

where c_0 is the speed of light in vacuum (see Chapter 5, p. 111).

More generally, following ideas outlined in [153] and [154], the basic equations for the electrostatic and electromagnetic forces may be written as

$$\boldsymbol{F}_{\text{es}} \;=\; k_{\text{es}}\frac{Q_1Q_2\boldsymbol{r}}{r^3}$$
$$\text{d}^2\boldsymbol{F}_{\text{em}} \;=\; \frac{k_{\text{em}}}{k^2}\frac{I_1\text{d}\boldsymbol{l}_1 \times (I_2\text{d}\boldsymbol{l}_2 \times \boldsymbol{r})}{r^3}$$

where the three constants k_{es}, k_{em} and k satisfy a more general Maxwell relation

$$k^2 \cdot k_{\text{es}}/k_{\text{em}} = c_0{}^2$$

The introduction of three constants is necessary in order to describe coherently the system of units and equations that are in common use in electromagnetic theory. These are, in addition to the SI, the esu (electrostatic unit) system, the emu (electromagnetic unit) system, the Gaussian system, and the system of atomic units. The constants k, k_{es}, and k_{em} have conventionally the values given in the following table.

	SI	esu	emu	Gaussian	atomic units	Notes
k	1	1	1	ζ cm s^{-1}	1	1
k_{es}	$1/4\pi\varepsilon_0$	1	ζ^2 cm^2 s^{-2}	1	1	2
k_{em}	$\mu_0/4\pi$	$(1/\zeta^2)$ s^2 cm^{-2}	1	1	α^2	3, 4

(1) ζ is the exact number $\zeta = c_0/(\text{cm s}^{-1}) = 29\,979\,245\,800$, see p. 135.
(2) $1/4\pi\varepsilon_0 = \zeta^2 \times 10^{-11}$ N A^{-2} m^2 s^{-2}
(3) $\mu_0/4\pi = 10^{-7}$ N A^{-2}
(4) α is the fine-structure constant with $\alpha^{-1} = 137.035\,999\,679(94)$ (see Chapter 5, p. 111).

This table can be used together with Section 7.4, p. 146 to transform expressions involving electromagnetic quantities between several systems. One sees that, particularly in the Gaussian system, $k = c_0$. This guarantees, together with the general relation of Maxwell equations given in Section 7.4, that the speed of light in vacuum comes out to be c_0 in the Gaussian system. Examples of transformations between the SI and the Gaussian system are also given in Section 7.4.

Additional remarks

(i) The esu system

In the esu system, the base units are the centimetre (cm), gram (g), and second (s) for length, mass and time, respectively. The franklin, symbol Fr (see footnote [1], below) for the electrostatic unit of charge may be introduced alternatively as a fourth base unit. Two particles with electric charges of one franklin, one centimetre apart in a vacuum, repel each other with a force of 1 dyn $= 1$ cm g s^{-2}. From this definition, one obtains the relation 1 Fr $= (10/\zeta)$ C (see Example 7, Section 7.1, p. 134). Since $k_{es} = 1$, from the general definition of the electrostatic force given above, the equation Fr $=$ erg$^{1/2}$ cm$^{1/2}$ is true in the esu system, where erg$^{1/2}$ cm$^{1/2}$ is the esu of charge. In this book the franklin is thus used for convenience as a name for the esu of charge.

(ii) The emu system

In the emu system, the base units are the centimetre (cm), gram (g), and second (s) for length, mass and time. The biot (symbol Bi) for the electromagnetic unit of electric current may be introduced alternatively as a fourth base unit. Two long wires separated by one centimetre with electric currents of one biot that flow in the same direction, repel each other in a vacuum with a lineic force (force per length) of 1 dyn/cm. From this definition one obtains the relation 1 Bi $= 10$ A. Since $k_{em} = 1$, from the general definition of the electromagnetic force given above, the equation Bi $=$ dyn$^{1/2}$ is true in the emu system, where dyn$^{1/2}$ is the emu of electric current. The biot has generally been used as a compact expression for the emu of electric current.

(iii) The Gaussian system

In the Gaussian system, the esu and emu systems are mixed. From the relation of the franklin and the biot to the SI units one readily obtains

$$1 \text{ Bi} = \zeta \text{ Fr s}^{-1}$$

Since $k_{es} = 1$ and $k_{em} = 1$, the value of the constant k is determined to be c_0 by the more general Maxwell relation given above. In treatises about relativity theory k is sometimes set to 1, which corresponds to a transformation of the time axis from a coordinate t to a coordinate $x = c_0 t$.

---

[1] The name "franklin", symbol Fr, for the esu of charge was suggested by Guggenheim [155], although it has not been widely adopted. The name "statcoulomb" has also been used for the esu of charge.

(iv) The atomic units system

In the system of atomic units [22] (see also Section 3.9.1, p. 94), charges are given in units of the elementary charge e, currents in units of $eE_h/\hbar$, where E_h is the hartree and $\hbar$ is the Planck constant divided by 2π, lengths are given in units of the bohr, a_0, the electric field strength is given in units of E_h/ea_0, the magnetic flux density in units of $\hbar/ea_0^2$. Conversion factors from these and other atomic units to the SI are included in Sections 3.9.1 and 7.2, p. 94 and 135. Thus, since conventionally $k = 1$, it follows that $k_{es} = 1$ in this system (see footnote [2]) and $k_{em} = \alpha^2$ (see footnote [3]).

(v) Non-rationalized quantities

The numerical constant 4π is introduced into the definitions of ε_0 and μ_0 because of the spherical symmetry involved in the equations defining $\boldsymbol{F}_{es}$ and $\mathrm{d}^2\boldsymbol{F}_{em}$ above; in this way its appearance is avoided in later equations, i.e., in the Maxwell equations. When factors of 4π are introduced in this way, as in the SI, the equations are also called "rationalized".

Furthermore, it is usual to include the factor 4π in the following quantities, when converting the electromagnetic equation from the SI system to the esu, emu, Gaussian and atomic units system:

$$
\begin{aligned}
D^{(ir)} &= 4\pi D \\
H^{(ir)} &= 4\pi H \\
\chi_e^{(ir)} &= \chi_e/4\pi \\
\chi^{(ir)} &= \chi/4\pi
\end{aligned}
$$

where the superscript (ir), for irrational, meaning non-rationalized, denotes the value of the corresponding quantity in a "non-rationalized" unit system as opposed to a system like the SI, which is "rationalized" in the sense described above.

The magnetic permeability μ is given as $\mu = \mu_r\, k_{em}\, 4\pi = \mu_r\, \mu_0$ in the SI, and as $\mu = \mu_r\, k_{em}$ in the non-rationalized unit system. μ_r is the dimensionless relative permeability and is defined in terms of the magnetic susceptibility in Section 7.4, p. 147. The electric permittivity ε is given as $\varepsilon = \varepsilon_r\, 4\pi/k_{es} = \varepsilon_r \cdot \varepsilon_0$ in the SI, and as $\varepsilon = \varepsilon_r/k_{es}$ in non-rationalized unit systems. ε_r is the dimensionless relative permittivity and is defined in terms of the electric susceptibility in Section 7.4, p. 146.

---

[2] Since $E_h = \dfrac{e^2}{4\pi\varepsilon_0 a_0}$, $E = \dfrac{1}{4\pi\varepsilon_0}\dfrac{Q}{r^2} = \dfrac{E_h a_0}{e^2}\dfrac{Q}{r^2} = \dfrac{E_h}{ea_0}\dfrac{(Q/e)}{(r/a_0)^2}$.

[3] Since the value of the speed of light in atomic units is the reciprocal of the fine-structure constant, α^{-1}, the condition $k^2\, k_{es}/k_{em} = c_0^2$ yields $k_{em} = \alpha^2$ in atomic units for $k = 1$.

7.4 TRANSFORMATION OF EQUATIONS OF ELECTROMAGNETIC THEORY BETWEEN THE ISQ(SI) AND GAUSSIAN FORMS

General relation	ISQ(SI)	Gaussian

Force between two localized charged particles in vacuum (Coulomb law):

$$\boldsymbol{F} = k_{es}Q_1 Q_2 \boldsymbol{r}/r^3 \qquad \boldsymbol{F} = Q_1 Q_2 \boldsymbol{r}/4\pi\varepsilon_0 r^3 \qquad \boldsymbol{F} = Q_1 Q_2 \boldsymbol{r}/r^3$$

Electrostatic potential around a localized charged particle in vacuum:

$$\phi = k_{es}Q/r \qquad \phi = Q/4\pi\varepsilon_0 r \qquad \phi = Q/r$$

Relation between electric field strength and electrostatic potential:

$$\boldsymbol{E} = -\boldsymbol{\nabla}\phi \qquad \boldsymbol{E} = -\boldsymbol{\nabla}\phi \qquad \boldsymbol{E} = -\boldsymbol{\nabla}\phi$$

Field due to a charge distribution ρ in vacuum (Gauss law):

$$\boldsymbol{\nabla}\cdot\boldsymbol{E} = 4\pi k_{es}\rho \qquad \boldsymbol{\nabla}\cdot\boldsymbol{E} = \rho/\varepsilon_0 \qquad \boldsymbol{\nabla}\cdot\boldsymbol{E} = 4\pi\rho$$

Electric dipole moment of a charge distribution:

$$\boldsymbol{p} = \int \rho\boldsymbol{r}\mathrm{d}V \qquad \boldsymbol{p} = \int \rho\boldsymbol{r}\mathrm{d}V \qquad \boldsymbol{p} = \int \rho\boldsymbol{r}\mathrm{d}V$$

Potential around a dipole in vacuum:

$$\phi = k_{es}\boldsymbol{p}\cdot\boldsymbol{r}/r^3 \qquad \phi = \boldsymbol{p}\cdot\boldsymbol{r}/4\pi\varepsilon_0 r^3 \qquad \phi = \boldsymbol{p}\cdot\boldsymbol{r}/r^3$$

Energy of a charge distribution in an electric field:

$$E = Q\phi - \boldsymbol{p}\cdot\boldsymbol{E} + \cdots \qquad E = Q\phi - \boldsymbol{p}\cdot\boldsymbol{E} + \cdots \qquad E = Q\phi - \boldsymbol{p}\cdot\boldsymbol{E} + \cdots$$

Electric dipole moment induced in an electric field:

$$\boldsymbol{p} = \alpha\boldsymbol{E} + \cdots \qquad \boldsymbol{p} = \alpha\boldsymbol{E} + \cdots \qquad \boldsymbol{p} = \alpha\boldsymbol{E} + \cdots$$

Dielectric polarization:

$$\boldsymbol{P} = \chi_e \boldsymbol{E}/4\pi k_{es} \qquad \boldsymbol{P} = \chi_e \varepsilon_0 \boldsymbol{E} \qquad \boldsymbol{P} = \chi_e^{(ir)}\boldsymbol{E}$$

Electric susceptibility and relative permittivity:

$$\varepsilon_r = 1 + \chi_e \qquad \varepsilon_r = 1 + \chi_e \qquad \varepsilon_r = 1 + 4\pi\chi_e^{(ir)}$$

Electric displacement[1]:

$$\boldsymbol{D} = \boldsymbol{E}/4\pi k_{es} + \boldsymbol{P} \qquad \boldsymbol{D} = \varepsilon_0 \boldsymbol{E} + \boldsymbol{P} \qquad \boldsymbol{D}^{(ir)} = \boldsymbol{E} + 4\pi\boldsymbol{P}$$

$$\boldsymbol{D} = \varepsilon_r \boldsymbol{E}/4\pi k_{es} \qquad \boldsymbol{D} = \varepsilon_0 \varepsilon_r \boldsymbol{E} \qquad \boldsymbol{D}^{(ir)} = \varepsilon_r \boldsymbol{E}$$

Capacitance of a parallel plate condenser, area A, separation d:

$$C = \varepsilon_r A/4\pi k_{es}d \qquad C = \varepsilon_0 \varepsilon_r A/d \qquad C = \varepsilon_r A/4\pi d$$

Force between two current elements in vacuum:

$$\mathrm{d}^2\boldsymbol{F} = \frac{k_{em}}{k^2}\frac{I_1\mathrm{d}\boldsymbol{l}_1 \times (I_2\mathrm{d}\boldsymbol{l}_2 \times \boldsymbol{r})}{r^3} \qquad \mathrm{d}^2\boldsymbol{F} = \frac{\mu_0}{4\pi}\frac{I_1\mathrm{d}\boldsymbol{l}_1 \times (I_2\mathrm{d}\boldsymbol{l}_2 \times \boldsymbol{r})}{r^3} \qquad \mathrm{d}^2\boldsymbol{F} = \frac{I_1\mathrm{d}\boldsymbol{l}_1 \times (I_2\mathrm{d}\boldsymbol{l}_2 \times \boldsymbol{r})}{c_0^2 r^3}$$

Magnetic vector potential due to a current element in vacuum:

$$\mathrm{d}\boldsymbol{A} = \frac{k_{em}}{k}I\mathrm{d}\boldsymbol{l}/r \qquad \mathrm{d}\boldsymbol{A} = \frac{\mu_0}{4\pi}I\mathrm{d}\boldsymbol{l}/r \qquad \mathrm{d}\boldsymbol{A} = I\mathrm{d}\boldsymbol{l}/c_0 r$$

Relation between magnetic flux density and magnetic vector potential:

$$\boldsymbol{B} = \boldsymbol{\nabla} \times \boldsymbol{A} \qquad \boldsymbol{B} = \boldsymbol{\nabla} \times \boldsymbol{A} \qquad \boldsymbol{B} = \boldsymbol{\nabla} \times \boldsymbol{A}$$

Magnetic flux density due to a current element in vacuum (Biot-Savart law):

$$\mathrm{d}\boldsymbol{B} = \frac{k_{em}}{k}I\mathrm{d}\boldsymbol{l} \times \boldsymbol{r}/r^3 \qquad \mathrm{d}\boldsymbol{B} = \frac{\mu_0}{4\pi}I\mathrm{d}\boldsymbol{l} \times \boldsymbol{r}/r^3 \qquad \mathrm{d}\boldsymbol{B} = I\mathrm{d}\boldsymbol{l} \times \boldsymbol{r}/c_0 r^3$$

(1) The second equation holds in isotropic media.

Magnetic flux density due to a current density $\boldsymbol{j}$ in vacuum (Ampère law):

$$\nabla \times \boldsymbol{B} = 4\pi \frac{k_{\mathrm{em}}}{k}\boldsymbol{j} \qquad \nabla \times \boldsymbol{B} = \mu_0 \boldsymbol{j} \qquad \nabla \times \boldsymbol{B} = 4\pi \boldsymbol{j}/c_0$$

Force on a current element in a magnetic flux density:

$$\mathrm{d}\boldsymbol{F} = I\mathrm{d}\boldsymbol{l} \times \boldsymbol{B}/k \qquad \mathrm{d}\boldsymbol{F} = I\mathrm{d}\boldsymbol{l} \times \boldsymbol{B} \qquad \mathrm{d}\boldsymbol{F} = I\mathrm{d}\boldsymbol{l} \times \boldsymbol{B}/c_0$$

Magnetic dipole of a current loop of area $\boldsymbol{A}$:

$$\boldsymbol{m} = I\boldsymbol{A}/k \qquad \boldsymbol{m} = I\boldsymbol{A} \qquad \boldsymbol{m} = I\boldsymbol{A}/c_0$$

Magnetic vector potential around a magnetic dipole in vacuum:

$$\boldsymbol{A} = k_{\mathrm{em}}\boldsymbol{m} \times \boldsymbol{r}/r^3 \qquad \boldsymbol{A} = \frac{\mu_0}{4\pi}\boldsymbol{m} \times \boldsymbol{r}/r^3 \qquad \boldsymbol{A} = \boldsymbol{m} \times \boldsymbol{r}/r^3$$

Energy of a magnetic dipole in a magnetic flux density:

$$E = -\boldsymbol{m} \cdot \boldsymbol{B} \qquad E = -\boldsymbol{m} \cdot \boldsymbol{B} \qquad E = -\boldsymbol{m} \cdot \boldsymbol{B}$$

Magnetic dipole induced by a magnetic flux density:

$$\boldsymbol{m} = \xi\boldsymbol{B} + \cdots \qquad \boldsymbol{m} = \xi\boldsymbol{B} + \cdots \qquad \boldsymbol{m} = \xi\boldsymbol{B} + \cdots$$

Magnetization:

$$\boldsymbol{M} = \chi\boldsymbol{H} \qquad \boldsymbol{M} = \chi\boldsymbol{H} \qquad \boldsymbol{M} = \chi^{(\mathrm{ir})}\boldsymbol{H}^{(\mathrm{ir})}$$

Magnetic susceptibility and relative permeability:

$$\mu_{\mathrm{r}} = 1 + \chi \qquad \mu_{\mathrm{r}} = 1 + \chi \qquad \mu_{\mathrm{r}} = 1 + 4\pi\chi^{(\mathrm{ir})}$$

Magnetic field strength[1]:

$$\boldsymbol{H} = \boldsymbol{B}/4\pi k_{\mathrm{em}} - \boldsymbol{M} \qquad \boldsymbol{H} = \boldsymbol{B}/\mu_0 - \boldsymbol{M} \qquad \boldsymbol{H}^{(\mathrm{ir})} = \boldsymbol{B} - 4\pi\boldsymbol{M}$$

$$\boldsymbol{H} = \boldsymbol{B}/4\pi\mu_r k_{\mathrm{em}} \qquad \boldsymbol{H} = \boldsymbol{B}/\mu_0\mu_r \qquad \boldsymbol{H}^{(\mathrm{ir})} = \boldsymbol{B}/\mu_r$$

Conductivity:

$$\boldsymbol{j} = \kappa\boldsymbol{E} \qquad \boldsymbol{j} = \kappa\boldsymbol{E} \qquad \boldsymbol{j} = \kappa\boldsymbol{E}$$

Self-inductance of a solenoid of volume V with n windings per length:

$$L = 4\pi\frac{k_{\mathrm{em}}}{k^2}\mu_{\mathrm{r}}n^2 V \qquad L = \mu_0\mu_{\mathrm{r}}n^2 V \qquad L = 4\pi\mu_{\mathrm{r}}n^2 V/c_0^2$$

Faraday induction law:

$$\nabla \times \boldsymbol{E} + \frac{1}{k}\frac{\partial \boldsymbol{B}}{\partial t} = \boldsymbol{0} \qquad \nabla \times \boldsymbol{E} + \partial \boldsymbol{B}/\partial t = \boldsymbol{0} \qquad \nabla \times \boldsymbol{E} + \frac{1}{c_0}\frac{\partial \boldsymbol{B}}{\partial t} = \boldsymbol{0}$$

Relation between the electric field strength and electromagnetic potentials:

$$\boldsymbol{E} = -\nabla\phi - \frac{1}{k}\frac{\partial \boldsymbol{A}}{\partial t} \qquad \boldsymbol{E} = -\nabla\phi - \partial \boldsymbol{A}/\partial t \qquad \boldsymbol{E} = -\nabla\phi - \frac{1}{c_0}\frac{\partial \boldsymbol{A}}{\partial t}$$

Maxwell equations:

$$\nabla \cdot \boldsymbol{D} = \rho \qquad \nabla \cdot \boldsymbol{D} = \rho \qquad \nabla \cdot \boldsymbol{D}^{(\mathrm{ir})} = 4\pi\rho$$

$$k\,\nabla \times \boldsymbol{H} - \partial D/\partial t = \boldsymbol{j} \qquad \nabla \times \boldsymbol{H} - \partial D/\partial t = \boldsymbol{j} \qquad \nabla \times \boldsymbol{H}^{(\mathrm{ir})} - \frac{1}{c_0}\frac{\partial \boldsymbol{D}^{(\mathrm{ir})}}{\partial t} = \frac{4\pi}{c_0}\boldsymbol{j}$$

$$k\,\nabla \times \boldsymbol{E} + \partial \boldsymbol{B}/\partial t = \boldsymbol{0} \qquad \nabla \times \boldsymbol{E} + \partial \boldsymbol{B}/\partial t = \boldsymbol{0} \qquad \nabla \times \boldsymbol{E} + \frac{1}{c_0}\frac{\partial \boldsymbol{B}}{\partial t} = \boldsymbol{0}$$

$$\nabla \cdot \boldsymbol{B} = 0 \qquad \nabla \cdot \boldsymbol{B} = 0 \qquad \nabla \cdot \boldsymbol{B} = 0$$

General relation	ISQ(SI)	Gaussian

Wave equations for the electromagnetic potentials ϕ and $\boldsymbol{A}$ (in Lorentz gauge):

$$\Delta\phi - \frac{k_{\text{em}}}{k^2\,k_{\text{es}}}\frac{\partial^2\phi}{\partial t^2} = -4\pi k_{\text{es}}\rho \qquad \Delta\phi - \varepsilon_0\mu_0\frac{\partial^2\phi}{\partial t^2} = -\rho/\varepsilon_0 \qquad \Delta\phi - \frac{1}{c_0{}^2}\frac{\partial^2\phi}{\partial t^2} = -4\pi\rho$$

$$\Delta\boldsymbol{A} - \frac{k_{\text{em}}}{k^2\,k_{\text{es}}}\frac{\partial^2\boldsymbol{A}}{\partial t^2} = -4\pi k_{\text{em}}\boldsymbol{j} \qquad \Delta\boldsymbol{A} - \varepsilon_0\mu_0\frac{\partial^2\boldsymbol{A}}{\partial t^2} = -\mu_0\boldsymbol{j} \qquad \Delta\boldsymbol{A} - \frac{1}{c_0{}^2}\frac{\partial^2\boldsymbol{A}}{\partial t^2} = -\frac{4\pi}{c_0}\boldsymbol{j}$$

$$\boldsymbol{\nabla A} + \frac{k_{\text{em}}}{k\,k_{\text{es}}}\frac{\partial\phi}{\partial t} = 0 \qquad \boldsymbol{\nabla A} + \varepsilon_0\mu_0\frac{\partial\phi}{\partial t} = 0 \qquad \boldsymbol{\nabla A} + \frac{1}{c_0}\frac{\partial\phi}{\partial t} = 0$$

Energy density of radiation:

$$U/V = (\boldsymbol{E}\cdot\boldsymbol{D} + \boldsymbol{B}\cdot\boldsymbol{H})/2 \qquad U/V = (\boldsymbol{E}\cdot\boldsymbol{D} + \boldsymbol{B}\cdot\boldsymbol{H})/2 \qquad U/V = \left(\boldsymbol{E}\cdot\boldsymbol{D}^{(\text{ir})} + \boldsymbol{B}\cdot\boldsymbol{H}^{(\text{ir})}\right)/8\pi$$

Rate of radiation energy flow (Poynting vector):

$$\boldsymbol{S} = k\,\boldsymbol{E}\times\boldsymbol{H} \qquad \boldsymbol{S} = \boldsymbol{E}\times\boldsymbol{H} \qquad \boldsymbol{S} = \frac{c_0}{4\pi}\boldsymbol{E}\times\boldsymbol{H}^{(\text{ir})}$$

Force on a moving charge Q with velocity $\boldsymbol{v}$ (Lorentz force):

$$\boldsymbol{F} = Q\left(\boldsymbol{E} + \boldsymbol{v}/k\times\boldsymbol{B}\right) \qquad \boldsymbol{F} = Q\left(\boldsymbol{E} + \boldsymbol{v}\times\boldsymbol{B}\right) \qquad \boldsymbol{F} = Q\left(\boldsymbol{E} + \boldsymbol{v}\times\boldsymbol{B}/c_0\right)$$

8 UNCERTAINTY

It is vital to report an uncertainty estimate along with a measurement. In brief, a report of a quantitative experimental (or theoretical) result should contain a statement about what the expected "best estimate" for the true result is, as well as a statement of the probable range of possible values specifying the uncertainty. The expected dispersion of the measured values arises from a number of sources. The contribution to the uncertainty from all these sources should be estimated as part of an uncertainty budget. The tables show some of the ways in which the contributions to the uncertainty budget can be estimated and combined and the examples below show how a full statement of the resulting combined uncertainty estimate can be presented. The present summary is based on [8].

8.1 REPORTING UNCERTAINTY FOR A SINGLE MEASURED QUANTITY

Uncertainties for a measured quantity X can be reported in several ways by giving the expected best estimate for the quantity together with a second quantity defining an uncertainty which corresponds to some probable range for the true quantity. We use the notation X for measured quantities, x for measurement results, and u_c for uncertainty.

Example 1. $m_s = 100.021\,47$ g with a combined standard uncertainty (i.e., estimated standard uncertainty) of $u_c = 0.35$ mg. Since it can be assumed that the possible estimated values of the standard are approximately normally distributed with approximate standard deviation u_c, the unknown value of the standard is believed to lie in the interval $m_s \pm u_c$ with a level of confidence of approximately 68 %.

Example 2. $m_s = (100.021\,47 \pm 0.000\,70)$ g, where the number following the symbol $\pm$ is the numerical value of an expanded uncertainty $U = k\,u_c$, with U determined from a combined standard uncertainty (i.e., estimated standard deviation) $u_c = 0.35$ mg and a coverage factor of $k = 2$. Since it can be assumed that the possible estimated values of the standard are approximately normally distributed with approximate standard deviation u_c, the unknown value of the standard is believed to lie within the interval defined by U with a level of confidence of 95 % (see Section 8.3, p. 154).

Example 3. $m_s = 100.021\,47(35)$ g, where the number in parentheses denotes the combined standard uncertainty $u_c = 0.35$ mg and is assumed to apply to the least significant digits.

Name	Symbol	Definition	Notes
Probability distributions			
probability distribution of x	$f(x)$	The probability density of the quantity having the value x; normal (Gaussian), rectangular, triangular, Student-t, etc.	
expected value of x	$E[x]$	$E[x] = \int x\, f(x)\mathrm{d}x$	
mean	μ	$E[x]$	
variance	σ^2	$E[(x-\mu)^2]$	
standard deviation	σ	$\sigma = +\sqrt{\sigma^2}$	
Statistics			
number of measurements	N		
mean	$\bar{x}$	$\bar{x} = \dfrac{1}{N}\sum\limits_{i=1}^{N} x_i$	
variance	$s^2(x)$	$s^2(x) = \dfrac{1}{N-1}\sum\limits_{i=1}^{N}\left(x_i - \bar{x}\right)^2$	1
standard deviation	$s(x)$	$s(x) = +\sqrt{s^2(x)}$	
standard deviation of the mean	$s(\bar{x})$	$s(\bar{x}) = s(x)/N$	

(1) This is an unbiased estimate and takes into account the removal of one degree of freedom because the spread is measured about the mean.

Name	Symbol	Definition	Notes
Uncertainties			
standard uncertainty of x_i	$u(x_i)$	Estimated by Type A or B approaches	2
Type A evaluation		Type A evaluation (of standard uncertainty) is a method of evaluation of a standard uncertainty by the statistical analysis of a series of observations	3
Type B evaluation		Type B evaluation (of standard uncertainty) is a method of evaluation of a standard uncertainty by means other than the statistical analysis of a series of observations (for instance, it is usually based on scientific judgement using all relevant information that is available)	4
relative standard uncertainty of x	$u_r(x_i)$	$u_r(x_i) = u(x_i)/\|x_i\| \quad (x_i \neq 0)$	
combined standard uncertainty of the measurement result y	$u_c(y)$	Estimated uncertainty in y from all the component measurement results $y = f(X_1, X_2, \cdots, X_m)$	5
relative combined standard uncertainty of y	$u_{c,r}(y)$	$u_{c,r}(y) = u_c(y)/\|y\| \quad (y \neq 0)$	
expanded uncertainty	U	$U = k\, u_c(y)$	6
coverage factor	k		6

(Notes continued)

(2) The uncertainty in the result of a measurement generally consists of several components which may be grouped into two categories according to the way in which their numerical value is estimated:

Type A: Those which are evaluated by statistical methods

Type B: Those which are evaluated by other means

A detailed report on uncertainty should consist of a complete list of the components, specifying for each the method used to obtain its numerical value. This provides the "uncertainty budget".

(3) Examples of type A evaluations are calculating the standard deviation of the mean of a series of independent observations; using the method of least squares to fit a function to data in order to estimate the parameters of the function and their standard deviations; and carrying out an analysis of variance (ANOVA) in order to identify and quantify random effects in certain kinds of measurements. These components of the uncertainty are characterized by the estimated variances s_i^2 (or the estimated "standard deviations" s_i) and the number of degrees of freedom ν_i. Where appropriate, the covariances should be given.

(4) This knowledge may include previous measurement data, general knowledge of the behavior and properties of relevant materials and instruments, manufacturer's specifications, data provided in calibration and other reports. These components of uncertainty should be characterized by quantities u_j^2, which may be considered approximations to the corresponding variances, the existence of which is assumed. The quantities u_j^2 may be treated like variances and the quantities u_j like standard deviations.

8.2 PROPAGATION OF UNCERTAINTY FOR UNCORRELATED MEASUREMENTS

For a discussion of the treatment of correlated uncertainties, see [8].

Measurement equation	Reported measurement result	Equation for the combined standard uncertainty	Notes
$Y = \sum_{i=1}^{N} a_i X_i$	$y = \sum_{i=1}^{N} a_i x_i$	$u_c(y) = \left(\sum_{i=1}^{N} a_i^2 u^2(x_i) \right)^{1/2}$	1
$Y = A X_1^{a_1} X_2^{a_2} \cdots X_N^{a_N}$	$y = A x_1^{a_1} x_2^{a_2} \cdots x_N^{a_N}$	$u_{c,r}(y) = \left(\sum_{i=1}^{N} a_i^2 u_r^2(x_i) \right)^{1/2}$	2

(1) This is sometimes known as addition in quadrature. The measurement equation is represented by a sum of quantities X_i multiplied by a constant a_i. The a_i are assumed to be known with certainty.

(2) The measurement equation is represented by a product of quantities X_i raised to powers $a_1, a_2, \cdots, a_N$ and multiplied by a constant A. A and the a_i are assumed to be known with certainty.

(Notes continued)

(4) (continued)

Examples of Type B evaluations:

(a) If it is reasonable to assume that the quantity, X, can be modeled by a normal probability distribution then lower and upper limits a_- and a_+ should be estimated such that the best estimated value of the input quantity is $x = (a_- + a_+)/2$ (i.e., the centre of the limits) and there is one chance out of two (i.e., a 50 % probability) that the value of the quantity lies in the interval a_- to a_+, then $u \approx 1.48(a_+ - a_-)/2$.

(b) If however the quantity, X, is better represented by a rectangular distribution then the lower and upper limits a_- and a_+ of the input quantity should be estimated such that the probability that the value lies in the interval a_- and a_+ is, for all practical purposes, 100 %. Provided that there is no contradictory information, treat the quantity as if it is equally probable for its value to lie anywhere within the interval a_- to a_+; that is, model it by a uniform (i.e., rectangular) probability distribution. The best estimate of the value of the quantity is then $x = (a_- + a_+)/2$ with the uncertainty $u = (a_+ - a_-)/\sqrt{3}$.

(5) The quantity, Y being measured, called the measurand, is not measured directly, but is determined from M other quantities $X_1, X_2, \cdots, X_M$ through a function f (sometimes called the measurement equation), $Y = f(X_1, X_2, \cdots, X_M)$. The quantities X_i include corrections (or correction factors), as well as quantities that take into account other sources of variability, such as different observers, instruments, samples, and laboratories etc. An estimate of the measured quantity Y, denoted $y = f(x_1, x_2, \cdots, x_M)$, is obtained from the measurement equation using *input estimates* $x_1, x_2, \cdots, x_M$ for the values of the M *input quantities*. A similar situation exists if other quantities are to be derived from measurements. The propagation of uncertainty is illustrated in the table above.

(6) Some commercial, industrial, and regulatory applications require a measure of uncertainty that defines an interval about a measurement result y within which the value of the measurand Y can be confidently asserted to lie. In these cases the *expanded uncertainty* U is used, and is obtained by multiplying the combined standard uncertainty, $u_c(y)$, by a *coverage factor* k. Thus $U = k\, u_c(y)$ and it is confidently believed that Y is greater than or equal to $y - U$, and is less than or equal to $y + U$, which is commonly written as $Y = y \pm U$. The value of the coverage factor k is chosen on the basis of the desired level of confidence to be associated with the interval defined by $U = k\, u_c$.

8.3 REPORTING UNCERTAINTIES IN TERMS OF CONFIDENCE INTERVALS

One can generally report the uncertainty by defining an interval, within which the quantity is expected to fall with a certain probability. This interval can be specified as $(y - U) \leq Y \leq (y + U)$.

Example   100.021 40 g $\leq m_\mathrm{s} \leq$ 100.021 54 g or $m_s = $ 100.021 47(7) g
or $m_\mathrm{s} = (100.021\,47 \pm 0.000\,07)$ g

When specifying such an interval, one should provide the confidence as the estimated probability with which the true quantity is expected to be in the given range (for example 60 %, 90 % or 95 %). This probability can be estimated using a variety of probability distributions appropriate for the measurement under consideration (Gaussian, Student-t, etc. [8,9]).

(Notes continued)

(6) (continued) Typically, k is in the range of two to three. When the normal distribution applies and u_c is a reliable estimate of the standard deviation of y, $U = 2u_\mathrm{c}$ (i.e., $k = 2$) defines an interval having a level of confidence of approximately 95 %, and $U = 3u_\mathrm{c}$ (i.e., $k = 3$) defines an interval having a level of confidence greater than 99 %.

9 ABBREVIATIONS AND ACRONYMS

Abbreviations and acronyms (words formed from the initial letters of groups of words that are frequently repeated) should be used sparingly. Unless they are well established (e.g. NMR, IR) they should always be defined once in any paper, and they should generally be avoided in titles and abstracts. Some acronyms have been accepted as common words, such as "laser" from the acronym LASER. Abbreviations used to denote physical quantities should if possible be replaced by the recommended symbol for the quantity (e.g. E_i rather than IP for ionization energy, see Section 2.5, p. 22; ρ rather than dens. for mass density, see Section 2.2, p. 14). For further recommendations concerning abbreviations see [156].

A list of frequently used abbreviations and acronyms is given here to help readers, but not necessarily to encourage their universal usage. In many cases an acronym can be found written in lower case letters and in capitals. In the list which follows only the most common usage is given. More extensive lists for different spectroscopic methods have been published by IUPAC [157, 158] and by Wendisch [159]; an extensive list for acronyms used in theoretical chemistry has been published by IUPAC [21].

A/D	analogue-to-digital
AA	atomic absorption
AAS	atomic absorption spectroscopy
ac	alternating current
ACM	adiabatic channel model
ACT	activated complex theory
AD	atom diffraction
ADC	analogue-to-digital converter
AES	Auger electron spectroscopy
AFM	atomic force microscopy
AIUPS	angle-integrated ultraviolet photoelectron spectroscopy
AM	amplitude modulated
amu	atomic mass unit
ANOVA	analysis of variance
AO	atomic orbital
APS	appearance potential spectroscopy
ARAES	angle-resolved Auger electron spectroscopy
ARPEFS	angle-resolved photoelectron fine structure
AS	Auger spectroscopy
ATR	attenuated total (internal) reflection
AU	astronomical unit
au	atomic unit
bcc	body-centred cubic
BET	Brunauer-Emmett-Teller (adsorption isotherm)
BIPM	Bureau International des Poids et Mesures
BIS	bremsstrahlung isochromat spectroscopy
BM	Bohr magneton (symbol: μ_B)
bp	boiling point
Btu	British thermal unit
CARS	coherent anti-Stokes Raman scattering
CAS	complete active space
CAS-SCF	complete active space - self consistent field
CAT	computer average of transients
CAWIA	Commission on Atomic Weights and Isotopic Abundances (formerly CIAAW)
CCA	coupled cluster approximation
CCC	critical coagulation concentration
CCD	coupled charge device
CCL	colour centre laser
CCU	Comité Consultatif d'Unités
ccp	cubic close packed
CD	circular dichroism
CEELS	characteristic electron energy-loss spectroscopy
CELS	characteristic energy-loss spectroscopy
CEPA	coupled electron pair approximation
CGPM	Conférence Générale des Poids et Mesures
cgs, CGS	centimetre-gram-second
CI	chemical ionization
CI	configuration interaction
CIAAW	Commission on Isotopic Abundances and Atomic Weights (now CAWIA)

CIDEP	chemically induced dynamic electron polarization
CIDNP	chemically induced dynamic nuclear polarization
CIMS	chemical ionization mass spectroscopy
CIPM	Comité International des Poids et Mesures
CIVR	collision induced vibrational relaxation
CIVR	classical intramolecular vibrational redistribution
CMA	cylindrical mass analyzer
CME	chemically modified electrode
CNDO	complete neglect of differential overlap
CODATA	Committee on Data for Science and Technology
COMAS	Concentration Modulation Absorption Spectroscopy
CPD	contact-potential difference
CRDS	cavity ring-down spectroscopy
CSRS	coherent Stokes-Raman scattering
CT	charge transfer
CVD	chemical vapor deposition
CW	continuous wave
D/A	digital-to-analogue
DAC	digital-to-analogue converter
D4WM	degenerate 4-wave mixing
DAPS	disappearance potential spectroscopy
dc, DC	direct current
DFG	difference frequency generation
DFT	density functional theory
DLVO	Derjaguin-Landau-Verwey-Overbeek
DME	dropping mercury electrode
DQMC	diffusion quantum Monte Carlo
DRIFTS	diffuse reflectance infrared Fourier transform spectroscopy
DSC	differential scanning calorimetry
DTA	differential thermal analysis
E1	elimination unimolecular
E2	elimination bimolecular
EAPFS	extended appearance potential fine structure
EC	electron capture
ECD	electron capture detector
ED	electron diffraction
EDA	electron donor-acceptor [complex]
EDX	energy-dispersive X-ray analysis
EELS	electron energy-loss spectroscopy
EH	electron holography
EI	electron impact ionization
EIS	electron impact spectroscopy
EIS	electrochemical impedance spectroscopy
EL	electroluminescence
ELDOR	electron-electron double resonance
ELEED	elastic low energy electron diffraction
emf	electromotive force
emu	electromagnetic unit

ENDOR	electron-nuclear double resonance
EPR	electron paramagnetic resonance
ESCA	electron spectroscopy for chemical applications (or analysis), see XPS
ESD	electron stimulated desorption
ESDIAD	electron stimulated desorption ion angular distribution
ESR	electron spin resonance
esu	electrostatic unit
ETS	electron transmission spectroscopy, electron tunneling spectroscopy
eu	entropy unit
EXAFS	extended X-ray absorption fine structure
EXAPS	electron excited X-ray appearance potential spectroscopy
EXELFS	extended electron energy-loss fine structure
FAB(MS)	fast atom bombardment (mass spectroscopy)
fcc	face-centred cubic
FD	field desorption
FEESP	field-emitted electron spin-polarization [spectroscopy]
FEM	field emission [electron] microscopy
FES	field emission spectroscopy
FFT	fast Fourier transform
FI	field ionization
FID	flame ionization detector
FID	free induction decay
FIM	field-ion microscopy
FIMS	field-ion mass spectroscopy
FIR	far-infrared
FM	frequency modulated
FPD	flame photometric detector
FSR	free spectral range
FT	Fourier transform
FTD	flame thermionic detector
FTIR	Fourier transform infrared
FWHM	full width at half maximum
GC	gas chromatography
GIXS	grazing-incidence X-ray scattering
GLC	gas-liquid chromatography
GM	Geiger-Müller
GTO	Gaussian-type orbital
GVB	generalized valence bond
hcp	hexagonal close-packed
HEED	high-energy electron diffraction
HEELS	high-energy electron energy-loss spectroscopy
HEIS	high-energy ion scattering
HF	Hartree-Fock
hfs	hyperfine structure (hyperfine splitting)
HMDE	hanging mercury drop electrode
HMO	Hückel molecular orbital
HOMO	highest occupied molecular orbital

HPLC	high-performance liquid chromatography
HREELS	high-resolution electron energy-loss spectroscopy
HTS	Hadamard transform spectroscopy
HWP	half-wave potential
I/O	input-output
IBA	ion beam analysis
IC	integrated circuit
ICISS	impact-collision ion scattering spectroscopy
ICR	ion cyclotron resonance
id	inner diameter
IEC	International Electrotechnical Commission
IEP	isoelectric point
IEPA	independent electron pair approximation
IETS	inelastic electron tunneling spectroscopy
ILEED	inelastic low-energy electron diffraction
INDO	incomplete neglect of differential overlap
INDOR	internuclear double resonance
INS	inelastic neutron scattering
INS	ion neutralization spectroscopy
IP	ionization potential (symbol: E_i)
IPES	inverse photoelectron spectroscopy
IPTS	International Practical Temperature Scale
IR	infrared
IRAS	infrared absorption spectroscopy
IRS	infrared spectroscopy
IS	ionization spectroscopy
ISO	International Organization for Standardization
ISQ	International System of Quantities
ISS	ion scattering spectroscopy
ITS	International Temperature Scale
IUPAC	International Union of Pure and Applied Chemistry
IUPAP	International Union of Pure and Applied Physics
KS	Kohn-Sham
L	ligand
L2TOFMS	laser desorption laser photoionization time-of-flight mass spectroscopy
LASER	light amplification by stimulated emission of radiation
LC	liquid chromatography
LCAO	linear combination of atomic orbitals
L-CCA	linear coupled-cluster approximation
LCMO	linear combination of molecular orbitals
LED	light-emitting diode
LEED	low-energy electron diffraction
LEELS	low-energy electron energy-loss spectroscopy
LEES	low-energy electron scattering
LEF	laser excitation fluorescence
LEIS	low-energy ion scattering
LEPD	low-energy positron diffraction

LET	linear energy transfer
LH	Lindemann-Hinshelwood [theory]
LID	laser induced desorption
LIDAR	light detection and ranging
LIF	laser induced fluorescence
LIGS	laser induced grating spectroscopy
LIMA	laser microprobe mass analysis
LIS	laser isotope separation
LMR	laser magnetic resonance
LUMO	lowest unoccupied molecular orbital
M	central metal
MAR	magic-angle rotation
MAS	magic-angle spinning
MASER	microwave amplification by stimulated emission of radiation
MBE	molecular beam epitaxy
MBGF	many-body Green's function
MBPT	many-body perturbation theory
MC	Monte Carlo
MCA	multichannel analyser
MCD	magnetic circular dichroism
MCS	multichannel scalar
MCSCF	multiconfiguration self-consistent field
MD	molecular dynamics
MDS	metastable deexcitation spectroscopy
MEED	medium-energy electron diffraction
MEIS	medium-energy ion scattering
MFM	magnetic force microscopy
MINDO	modified incomplete neglect of differential overlap
MIR	mid-infrared
MKS	metre-kilogram-second
MKSA	metre-kilogram-second-ampere
MM	molecular mechanics
MO	molecular orbital
MOCVD	metal-organic chemical vapor deposition
MOMBE	metal-organic molecular beam epitaxy
MORD	magnetic optical rotatory dispersion
MOS	metal oxide semiconductor
mp	melting point
MPI	multiphoton ionization
MPPT	Møller-Plesset perturbation theory
MP-SCF	Møller-Plesset self-consistent field
MRD	magnetic rotatory dispersion
MRI	magnetic resonance imaging
MS	mass spectroscopy
MW	microwave
MW	molecular weight (symbol: M_r)
NAA	neutron activation analysis
NCE	normal calomel electrode
Nd:YAG	Nd doped YAG

161

NETD	noise equivalent temperature difference
NEXAFS	near edge X-ray absorption fine structure
NIR	near-infrared
NIR	non-ionizing radiation
NMA	nuclear microanalysis
NMR	nuclear magnetic resonance
NOE	nuclear Overhauser effect
NQR	nuclear quadrupole resonance
NTP	normal temperature and pressure
od	outside diameter
ODMR	optically detected magnetic resonance
OGS	opto-galvanic spectroscopy
ORD	optical rotatory dispersion
PAS	photoacoustic spectroscopy
PC	paper chromatography
PD	photoelectron diffraction
PDG	Particle Data Group
PED	photoelectron diffraction
PEH	photoelectron holography
PES	photoelectron spectroscopy
PIES	Penning ionization electron spectroscopy, see PIS
PIPECO	photoion-photoelectron coincidence [spectroscopy]
PIPN	periodically poled lithium niobate
PIS	Penning ionization (electron) spectroscopy
PMT	photomultiplier tube
ppb	part per billion
pphm	part per hundred million
ppm	part per million
PPP	Pariser-Parr-Pople
PS	see PES
PSD	photon stimulated desorption
pzc	point of zero charge
QET	quasi equilibrium theory
QMB	quartz microbalance
QMC	quantum Monte Carlo
QMS	quadrupole mass spectrometer
RADAR	radiowave detection and ranging
RBS	Rutherford (ion) back scattering
RD	rotatory dispersion
RDE	rotating disc electrode
RDF	radial distribution function
REM	reflection electron microscopy
REMPI	resonance enhanced multiphoton ionization
RF	radio frequency
RHEED	reflection high-energy electron diffraction
RHF	restricted Hartree-Fock
RIMS	resonant ionization mass spectroscopy

RKR	Rydberg-Klein-Rees [potential]
rms	root-mean-square
RRK	Rice-Ramsperger-Kassel [theory]
RRKM	Rice-Ramsperger-Kassel-Marcus [theory]
RRS	resonance Raman spectroscopy
RS	Raman spectroscopy
RSPT	Rayleigh-Schrödinger perturbation theory
S	singlet
SACM	statistical adiabatic channel model
SAM	scanning Auger microscopy
SBS	stimulated Brillouin scattering
SBZ	surface Brillouin zone
SCE	saturated calomel electrode
SCF	self-consistent field
SDCI	singly and doubly excited configuration interaction
S_E	substitution electrophilic
SEELFS	surface extended energy-loss fine structure
SEFT	spin-echo Fourier transform
SEM	scanning [reflection] electron microscopy
SEP	stimulated emission pumping
SERS	surface-enhanced Raman spectroscopy
SES	secondary electron spectroscopy
SESCA	scanning electron spectroscopy for chemical applications
SEXAFS	surface extended X-ray absorption fine structure
SF	spontaneous fission
SFG	sum-frequency generation
SHE	standard hydrogen electrode
SHG	second-harmonic generation
SI	Système International d'Unités
SIMS	secondary ion mass spectroscopy
SMOKE	surface magneto-optic Kerr effect
S_N1	substitution nucleophilic unimolecular
S_N2	substitution nucleophilic bimolecular
S_Ni	substitution nucleophilic intramolecular
SOC	spin-orbit coupling
SOR	synchrotron orbital radiation
SPIES	surface Penning ionization electron spectroscopy
SPLEED	spin-polarized low-energy electron diffraction
SPM	scanning probe microscopy
SR	synchrotron radiation
SRS	synchrotron radiation source
SSIMS	static secondary ion mass spectroscopy
STEM	scanning transmission [electron] microscopy
STM	scanning tunnelling (electron) microscopy
STO	Slater-type orbital
STP	standard temperature and pressure
SVLF	single vibronic level fluorescence

T	triplet
TCC	thermal conductivity cell
TCD	thermal conductivity detector
TCF	time correlation function
TDL	tuneable diode laser
TDMS	tandem quadrupole mass spectroscopy
TDS	thermal desorption spectroscopy
TED	transmission electron diffraction
TEM	transmission electron microscopy
TG	thermogravimetry
TGA	thermogravimetric analysis
THEED	transmission high-energy electron diffraction
tlc	thin layer chromatography
TOF	time-of-flight [analysis]
TPD	temperature programmed desorption
TPR	temperature programmed reaction
TR3	time-resolved resonance Raman scattering
TST	transition state theory
UHF	ultra high frequency
UHF	unrestricted Hartree-Fock
UHV	ultra high vacuum
UP[E]S	ultraviolet photoelectron spectroscopy
UV	ultraviolet
VB	valence bond
VCD	vibrational circular dichroism
VEELS	vibrational electron energy-loss spectroscopy
VHF	very high frequency
VIS	visible
VLEED	very-low-energy electron diffraction
VLSI	very large scale integration
VPC	vapor-phase chromatography
VSEPR	valence shell electron pair repulsion
VUV	vacuum ultraviolet
WFC	work function change
X	halogen
XAFS	X-ray absorption fine structure
XANES	X-ray absorption near-edge structure [spectroscopy]
XAPS	X-ray appearance potential spectroscopy
XPD	X-ray photoelectron diffraction
XPES	X-ray photoelectron spectroscopy
XPS	X-ray photoelectron spectroscopy
XRD	X-ray diffraction
XSW	X-ray standing wave
YAG	yttrium aluminium garnet
ZPE	zero point energy

10 REFERENCES

10.1 PRIMARY SOURCES

[1] Manual of Symbols and Terminology for Physicochemical Quantities and Units
   (a) 1st ed., M. L. McGlashan. *Pure Appl. Chem.*, 21:1−38, 1970.
   (b) 2nd ed., M. A. Paul. Butterworths, London, 1975.
   (c) 3rd ed., D. H. Whiffen. *Pure Appl. Chem.*, 51:1−36, 1979.
   (d) D. H. Whiffen. Appendix I−Definitions of Activities and Related Quantities. *Pure Appl. Chem.*, 51:37−41, 1979.
   (e) D. H. Everett. Appendix II−Definitions, Terminology and Symbols in Colloid and Surface Chemistry, Part I. *Pure Appl. Chem.*, 31:577−638, 1972.
   (f) J. Lyklema and H. van Olphen. Appendix II−Definitions, Terminology and Symbols in Colloid and Surface Chemistry, Part 1.13: Selected Definitions, Terminology and Symbols for Rheological Properties. *Pure Appl. Chem.*, 51:1213−1218, 1979.
   (g) M. Kerker and J. P. Kratochvil. Appendix II−Definitions, Terminology and Symbols in Colloid and Surface Chemistry, Part 1.14: Light Scattering. *Pure Appl. Chem.*, 55:931−941, 1983.
   (h) R. L. Burwell Jr. Part II: Heterogeneous Catalysis. *Pure Appl. Chem.*, 46:71−90, 1976.
   (i) R. Parsons. Appendix III−Electrochemical Nomenclature. *Pure Appl. Chem.*, 37:499−516, 1974.
   (j) J. D. Cox. Appendix IV−Notation for States and Processes, Significance of the Word "Standard" in Chemical Thermodynamics, and Remarks on Commonly Tabulated Forms of Thermodynamic Functions. *Pure Appl. Chem.*, 54:1239−1250, 1982.
   (k) K. J. Laidler. Appendix V−Symbolism and Terminology in Chemical Kinetics. *Pure Appl. Chem.*, 53:753−771, 1981.
[2] (a) I. Mills, T. Cvitaš, K. Homann, N. Kallay and K. Kuchitsu. *Quantities, Units and Symbols in Physical Chemistry.* Blackwell Science, Oxford, 1st edition, 1988.
   (b) I. Mills, T. Cvitaš, K. Homann, N. Kallay and K. Kuchitsu. *Quantities, Units and Symbols in Physical Chemistry.* Blackwell Science, Oxford, 2nd edition, 1993.
   (c) *Nomenklaturniye Pravila IUPAC po Khimii, Vol. 6, Fizicheskaya Khimiya,* Nacionalnii Komitet Sovetskih Khimikov, Moscow, 1988.
   (d) M. Riedel. *A Fizikai-kémiai Definiciók és Jelölések.* Tankönyvkiadó, Budapest, 1990.
   (e) K. Kuchitsu. *Butsurikagaku de Mochiirareru Ryo, Tan-i, Kigo.* Kodansha, Tokyo, 1991.
   (f) K.-H. Homann, M. Hausmann. *Größen, Einheiten und Symbole in der Physikalischen Chemie.* VCH, Weinheim, 1995.
   (g) D. I. Marchidan. *Mărimi, Unităţi şi Simboluri în Chimia Fizică.* Editura Academiei Române, Bucharest, 1996.
   (h) A. P. Masiá, J. M. Guil, J. E. Herrero, A. R. Paniego. *Magnitudes, Unidades y Símbolos en Química Física.* Fundació Ramón Areces & Real Sociedad Española de Química, Madrid, 1999.
   (i) J. M. Costa. *Magnituds, Unitats i Símbols en Química Física.* Institut d'Estudis Catalans, Barcelona, 2004.
[3] Bureau International des Poids et Mesures. *Le Système International d'Unités (SI).* 8th French and English Edition, BIPM, Sèvres, 2006.
[4] E. R. Cohen and P. Giacomo. *Symbols, Units, Nomenclature and Fundamental Constants in Physics. 1987 Revision*, Document I.U.P.A.P.−25 (IUPAP-SUNAMCO 87−1) also published in: *Physica*, 146A:1−67, 1987.

[5] International Standards, ISO
International Organization for Standardization, Geneva, Switzerland
   (a) ISO 31-0:   1992,   Quantities and units – Part 0:   General principles
      Amendment 1: 1998, Amendment 2: 2005
   (b) ISO 80000-3: 2006,   Quantities and units – Part 3:   Space and time
   (c) ISO 80000-4: 2006,   Quantities and units – Part 4:   Mechanics
   (d) ISO 31-4:   1992,   Quantities and units – Part 4:   Heat
      Amendment 1: 1998
   (e) ISO 31-5:   1992,   Quantities and units – Part 5:   Electricity and magnetism
      Amendment 1: 1998
   (f) ISO 31-6:   1992,   Quantities and units – Part 6:   Light and related electromagnetic radiations
      Amendment 1: 1998
   (g) ISO 31-7:   1992,   Quantities and units – Part 7:   Acoustics
      Amendment 1: 1998
   (h) ISO 31-8:   1992,   Quantities and units – Part 8:   Physical chemistry and molecular physics
      Amendment 1: 1998
   (i) ISO 31-9:   1992,   Quantities and units – Part 9:   Atomic and nuclear physics
      Amendment 1: 1998
   (j) ISO 31-10:   1992,   Quantities and units – Part 10:   Nuclear reactions and ionizing radiations
      Amendment 1: 1998
   (k) ISO 31-11:   1992,   Quantities and units – Part 11:   Mathematical signs and symbols for use in the physical sciences and technology
   (m) ISO 31-12:   1992,   Quantities and units – Part 12:   Characteristic numbers
      Amendment 1: 1998
   (n) ISO 31-13:   1992,   Quantities and units – Part 13:   Solid state physics
      Amendment 1: 1998
The different parts of ISO 31 are successively being replaced with ISO/IEC 80000.

[6] ISO 1000: 1992,   SI Units and recommendations for the use of their multiples and of certain other units
   Amendment 1: 1998

All the standards listed here ([5] and [6]) are jointly reproduced in the ISO Standards Handbook 2, *Quantities and units*, ISO, Geneva, 1993.

[7] International Standard, IEC International Electrotechnical Commission, Geneva.
IEC 60027–2, 2005.
[8] Guide to the Expression of Uncertainty in Measurement (GUM), BIPM, IEC, IFCC, ISO, IUPAC, IUPAP, OIML, International Organization for Standardization, Geneva.
1st edition, 1993; corrected and reprinted, 1995.
[9] International Vocabulary of Metrology – Basic and General Concepts and Associated Terms (VIM), 3rd edition, 2007.

10.2 SECONDARY SOURCES

[10] M. Quack. Commission I.1 at the IUPAC General Assembly 1995. Summary Minutes. *Chem. Int.*, 20:12, 1998.

[11] E. A. Guggenheim. Units and Dimensions. *Phil. Mag.*, 33:479–496, 1942.

[12] J. de Boer. On the History of Quantity Calculus and the International System. *Metrologia*, 31:405–429, 1994/95.

[13] I. M. Mills. The Language of Science. *Metrologia*, 34:101–109, 1997.

[14] J. C. Rigg, S. S. Brown, R. Dybkaer, and H. Olesen, editors. *Compendium of Terminology and Nomenclature of Properties in Clinical Laboratory Sciences.* Blackwell, Oxford, 1995.

[15] J. Rigaudy and S. P. Klesney. *IUPAC Nomenclature of Organic Chemistry, Sections A, B, C, D, E, F, and H.* Pergamon Press, Oxford, 4th edition, 1979.

[16] A. D. McNaught and A. Wilkinson. *Compendium of Chemical Terminology – The Gold Book,* 2nd edition. Blackwell, Oxford, 1997.

[17] P. Müller. Glossary of Terms Used in Physical Organic Chemistry. *Pure Appl. Chem.*, 66:1077–1184, 1994.

[18] A. D. Jenkins, P. Kratochvíl, R. F. T. Stepto, and U. W. Suter. Glossary of Basic Terms in Polymer Science. *Pure Appl. Chem.*, 68:2287–2311, 1996.

[19] L. D. Landau and E. M. Lifshitz, editors. *Course of Theoretical Physics: Mechanics.* Butterworth-Heinemann, Oxford, 1998.

[20] J. E. Boggs. Guidelines for Presentation of Methodological Choices in the Publication of Computational Results. A. *Ab Initio* Electronic Structure Calculations. *Pure Appl. Chem.*, 70:1015–1018, 1998.

[21] R. D. Brown, J. E. Boggs, R. Hilderbrandt, K. Lim, I. M. Mills, E. Nikitin, and M. H. Palmer. Acronyms Used in Theoretical Chemistry. *Pure Appl. Chem.*, 68:387–456, 1996.

[22] D. H. Whiffen. Expression of Results in Quantum Chemistry. *Pure Appl. Chem.*, 50:75–79, 1978.

[23] P. J. Mohr, B. N. Taylor, and D. B. Newell. CODATA Recommended Values of the Fundamental Physical Constants: 2006, *to be published*. The 2006 CODATA recommended values are available at `http://physics.nist.gov/constants`.

[24] J. Mullay. Estimation of Atomic and Group Electronegativities. *Struct. Bonding (Berlin)*, 66:1–25, 1987.

[25] J. M. Brown, R. J. Buenker, A. Carrington, C. di Lauro, R. N. Dixon, R. W. Field, J. T. Hougen, W. Hüttner, K. Kuchitsu, M. Mehring, A. J. Merer, T. A. Miller, M. Quack, D. A. Ramsey, L. Veseth, and R. N. Zare. Remarks on the Signs of g-Factors in Atomic and Molecular Zeeman Spectroscopy. *Mol. Phys.*, 98:1597–1601, 2000.

[26] M. Quack and J. Stohner. Combined Multidimensional Anharmonic and Parity Violating Effects in CDBrClF. *J. Chem. Phys.*, 119:11228–11240, 2003.

[27] F. A. Jenkins. Report of Subcommittee f (Notation for the Spectra of Diatomic Molecules). *J. Opt. Soc. Amer.*, 43:425–426, 1953.

[28] R. S. Mulliken. Report on Notation for the Spectra of Polyatomic Molecules. *J. Chem. Phys.*, 23:1997–2011, 1955. (Erratum) *J. Chem. Phys.*, 24:1118, 1956.

[29] G. Herzberg. *Molecular Spectra and Molecular Structure Vol. II. Infrared and Raman Spectra of Polyatomic Molecules.* Van Nostrand, Princeton, 1946.

[30] G. Herzberg. *Molecular Spectra and Molecular Structure Vol. I. Spectra of Diatomic Molecules.* Van Nostrand, Princeton, 1950.

[31] G. Herzberg. *Molecular Spectra and Molecular Structure Vol. III. Electronic Spectra and Electronic Structure of Polyatomic Molecules.* Van Nostrand, Princeton, 1966.

[32] E. D. Becker. Recommendations for the Presentation of Infrared Absorption Spectra in Data Collections – A. Condensed Phases. *Pure Appl. Chem.*, 50:231–236, 1978.

[33] E. D. Becker, J. R. Durig, W. C. Harris, and G. J. Rosasco. Presentation of Raman Spectra in Data Collections. *Pure Appl. Chem.*, 53:1879–1885, 1981.

[34] Physical Chemistry Division, Commission on Molecular Structure and Spectroscopy. Recommendations for the Presentation of NMR Data for Publication in Chemical Journals. *Pure Appl. Chem.*, 29:625–628, 1972.

[35] Physical Chemistry Division, Commission on Molecular Structure and Spectroscopy. Presentation of NMR Data for Publication in Chemical Journals – B. Conventions Relating to Spectra from Nuclei other than Protons. *Pure Appl. Chem.*, 45:217–219, 1976.

[36] Physical Chemistry Division, Commission on Molecular Structure and Spectroscopy. Nomenclature and Spectral Presentation in Electron Spectroscopy Resulting from Excitation by Photons. *Pure Appl. Chem.*, 45:221–224, 1976.

[37] Physical Chemistry Division, Commission on Molecular Structure and Spectroscopy. Nomenclature and Conventions for Reporting Mössbauer Spectroscopic Data. *Pure Appl. Chem.*, 45:211–216, 1976.

[38] J. H. Beynon. Recommendations for Symbolism and Nomenclature for Mass Spectroscopy. *Pure Appl. Chem.*, 50:65–73, 1978.

[39] C. J. H. Schutte, J. E. Bertie, P. R. Bunker, J. T. Hougen, I. M. Mills, J. K. G. Watson, and B. P. Winnewisser. Notations and Conventions in Molecular Spectroscopy: Part 1. General Spectroscopic Notation. *Pure Appl. Chem.*, 69:1633–1639, 1997.

[40] C. J. H. Schutte, J. E. Bertie, P. R. Bunker, J. T. Hougen, I. M. Mills, J. K. G. Watson, and B. P. Winnewisser. Notations and Conventions in Molecular Spectroscopy: Part 2. Symmetry Notation. *Pure Appl. Chem.*, 69:1641–1649, 1997.

[41] P. R. Bunker, C. J. H. Schutte, J. T. Hougen, I. M. Mills, J. K. G. Watson, and B. P. Winnewisser. Notations and Conventions in Molecular Spectroscopy: Part 3. Permutation and Permutation-Inversion Symmetry Notation. *Pure Appl. Chem.*, 69:1651–1657, 1997.

[42] R. K. Harris, E. D. Becker, S. M. Cabral de Menezes, R. Goodfellow, and P. Granger. Parameters and Symbols for Use in Nuclear Magnetic Resonance. *Pure Appl. Chem.*, 69:2489–2495, 1997.

[43] J. L. Markley, A. Bax, Y. Arata, C. W. Hilbers, R. Kaptein, B. D. Sykes, P. E. Wright, and K. Wüthrich. Recommendations for the Presentation of NMR Structures of Proteins and Nucleic Acids. *Pure Appl. Chem.*, 70:117–142, 1998.

[44] J. K. G. Watson. Aspects of Quartic and Sextic Centrifugal Effects on Rotational Energy Levels. In J. R. Durig, editor, *Vibrational Spectra and Structure, Vol. 6*, pages 1–89, Amsterdam, 1977. Elsevier.

[45] G. Graner, E. Hirota, T. Iijima, K. Kuchitsu, D. A. Ramsay, J. Vogt, and N. Vogt. Structure Data of Free Polyatomic Molecules. In K. Kuchitsu, editor, *Landolt-Börnstein, New Series, II/25C*, pages 7–10, Berlin, 2000. Springer.

[46] Y. Morino and T. Shimanouchi. Definition and Symbolism of Molecular Force Constants. *Pure Appl. Chem.*, 50:1707–1713, 1978.

[47] R. K. Harris, E. D. Becker, S. M. Cabral de Menezes, R. Goodfellow, and P. Granger. NMR Nomenclature. Nuclear Spin Properties and Conventions for Chemical Shifts. *Pure Appl. Chem.*, 73:1795–1818, 2001.

[48] P. R. Bunker and P. Jensen. *Molecular Symmetry and Spectroscopy*, 2nd edition. NRC Research Press, Ottawa, 1998.

[49] J. M. Brown, J. T. Hougen, K.-P. Huber, J. W. C. Johns, I. Kopp, H. Lefebvre-Brion, A. J. Merer, D. A. Ramsay, J. Rostas, and R. N. Zare. The Labeling of Parity Doublet Levels in Linear Molecules. *J. Mol. Spectrosc.*, 55:500–503, 1975.

[50] M. H. Alexander, P. Andresen, R. Bacis, R. Bersohn, F. J. Comes, P. J. Dagdigian, R. N. Dixon, R. W. Field, G. W. Flynn, K.-H. Gericke, E. R. Grant, B. J. Howard, J. R. Huber, D. S. King, J. L. Kinsey, K. Kleinermanns, K. Kuchitsu, A. C. Luntz, A. J. McCaffery, B. Pouilly, H. Reisler, S. Rosenwaks, E. W. Rothe, M. Shapiro, J. P. Simons, R. Vasudev, J. R. Wiesenfeld, C. Wittig, and R. N. Zare. A Nomenclature for Λ-doublet Levels in Rotating Linear Molecules. *J. Chem. Phys.*, 89:1749–1753, 1988.

[51] M. Hamermesh. *Group Theory and Its Application to Physical Problems*. Addison-Wesley, Reading, 1962.

[52] H. C. Longuet-Higgins. The Symmetry Groups of Non-Rigid Molecules. *Mol. Phys.*, 6:445–460, 1963.

[53] I. M. Mills and M. Quack. Introduction to 'The Symmetry Groups of Non-Rigid Molecules' by H. C. Longuet-Higgins. *Mol. Phys.*, 100:9–10, 2002.

[54] M. Quack. Detailed Symmetry Selection Rules for Reactive Collisions. *Mol. Phys.*, 34:477–504, 1977.

[55] J. C. D. Brand, J. H. Callomon, K. K. Innes, J. Jortner, S. Leach, D. H. Levy, A. J. Merer, I. M. Mills, C. B. Moore, C. S. Parmenter, D. A. Ramsay, K. N. Rao, E. W. Schlag, J. K. G. Watson, and R. N. Zare. The Vibrational Numbering of Bands in the Spectra of Polyatomic Molecules. *J. Mol. Spectrosc.*, 99:482–483, 1983.

[56] M. Terazima, N. Hirota, S. E. Braslavsky, A. Mandelis, S. E. Bialkowski, G. J. Diebold, R. J. D. Miller, D. Fournier, R. A. Palmer, and A. Tam. Quantities, Terminology, and Symbols in Photothermal and Related Spectroscopies. *Pure Appl. Chem.*, 76:1083–1118, 2004.

[57] N. Sheppard, H. A. Willis, and J. C. Rigg. Names, Symbols, Definitions and Units of Quantities in Optical Spectroscopy. *Pure Appl. Chem.*, 57:105–120, 1985.

[58] V. A. Fassel. Nomenclature, Symbols, Units and their Usage in Spectrochemical Analysis. I. General Atomic Emission Spectroscopy. *Pure Appl. Chem.*, 30:651–679, 1972.

[59] W. H. Melhuish. Nomenclature, Symbols, Units and their Usage in Spectrochemical Analysis. VI: Molecular Luminescence Spectroscopy. *Pure Appl. Chem.*, 56:231–245, 1984.

[60] J. W. Verhoeven. Glossary of Terms Used in Photochemistry. *Pure Appl. Chem.*, 68:2223–2286, 1996.

[61] A. A. Lamola and M. S. Wrighton. Recommended Standards for Reporting Photochemical Data. *Pure Appl. Chem.*, 56:939–944, 1984.

[62] M. Quack. Spectra and Dynamics of Coupled Vibrations in Polyatomic Molecules. *Annu. Rev. Phys. Chem.*, 41:839–874, 1990.

[63] A. G. Maki and J. S. Wells. Wavenumber Calibration Tables from Heterodyne Frequency Measurements. *NIST Special Publication 821, U.S. Department of Commerce*, 1991.

[64] L. A. Pugh and K. N. Rao. Intensities from Infrared Spectra. In K. N. Rao, editor, *Molecular Spectroscopy: Modern Research, Vol. 2*, New York, 1978. Academic Press.

[65] M. A. Smith, C. P. Rinsland, B. Fridovich, and K. N. Rao. Intensities and Collision Broadening Parameters from Infrared Spectra. In K. N. Rao, editor, *Molecular Spectroscopy: Modern Research, Vol. 3*, New York, 1985. Academic Press.

[66] M. Quack. Multiphoton Excitation. In P. von Ragué Schleyer, N. Allinger, T. Clark, J. Gasteiger, P. A. Kollman, H. F. Schaefer III, and P. R. Schreiner, editors, *Encyclopedia of Computational Chemistry, Vol. 3*, pages 1775–1791, Chichester, UK, 1998. John Wiley and Sons.

[67] J. E. Bertie, S. L. Zhang, H. H. Eysel, S. Baluja, and M. K. Ahmed. Infrared Intensities of Liquids XI: Infrared Refractive Indices from 8000 to 2 cm^{-1}, Absolute Integrated Intensities, and Dipole Moment Derivatives of Methanol at 25 °C. *Appl. Spec.*, 47:1100–1114, 1993.

[68] J. E. Bertie and C. Dale Keefe. Comparison of Infrared Absorption Intensities of Benzene in the Liquid and Gas Phases. *J. Chem. Phys.*, 101:4610–4616, 1994.

[69] S. R. Polo and M. K. Wilson. Infrared Intensities in Liquid and Gas Phases. *J. Chem. Phys.*, 23:2376–2377, 1955.

[70] Th. Hahn, editor. *International Tables for Crystallography, Vol. A: Space-Group Symmetry*. Reidel, Dordrecht, 2nd edition, 1983.

[71] S. Trasatti and R. Parsons. Interphases in Systems of Conducting Phases. *Pure Appl. Chem.*, 58:437–454, 1986.

[72] T. Cvitaš. Quantities Describing Compositions of Mixtures. *Metrologia*, 33:35–39, 1996.

[73] H. P. Lehmann, X. Fuentes-Arderiu, and L. F. Bertello. Glossary of Terms in Quantities and Units in Clinical Chemistry. *Pure Appl. Chem.*, 68:957–1000, 1996.

[74] N. G. Connelly, T. Damhus, R. M. Hartshorn, and A. T. Hutton, editors. *Nomenclature of Inorganic Chemistry – IUPAC Recommendations 2005*. The Royal Society of Chemistry, Cambridge, 2005.

[75] W. H. Koppenol. Names for Muonium and Hydrogen Atoms and their Ions. *Pure Appl. Chem.*, 73:377–380, 2001.

[76] W. H. Powell. Revised Nomenclature for Radicals, Ions, Radical Ions and Related Species. *Pure Appl. Chem.*, 65:1357–1455, 1993.

[77] W. H. Koppenol. Names for Inorganic Radicals. *Pure Appl. Chem.*, 72:437–446, 2000.

[78] J. Brecher. Graphical Representation of Stereochemical Configuration. *Pure Appl. Chem.*, 78:1897–1970, 2006.

[79] G. P. Moss. Basic Terminology of Stereochemistry. *Pure Appl. Chem.*, 68:2193–2222, 1996.

[80] R. A. Alberty. Chemical Equations are Actually Matrix Equations. *J. Chem. Educ.*, 68:984, 1991.

[81] M. B. Ewing, T. H. Lilley, G. M. Olofsson, M. T. Rätzsch, and G. Somsen. Standard Quantities in Chemical Thermodynamics. Fugacities, Activities, and Equilibrium Constants for Pure and Mixed Phases. *Pure Appl. Chem.*, 66:533–552, 1994.

[82] H. Preston-Thomas. The International Temperature Scale of 1990 (ITS-90). *Metrologia*, 27:3–10, 1990. BIPM "Echelle International de Température" de 1990 (EIT90), ISO 31-4 1992 (E).

[83] R. N. Goldberg and R. D. Weir. Conversion of Temperatures and Thermodynamic Properties to the Basis of the International Temperature Scale of 1990. *Pure Appl. Chem.*, 64:1545–1562, 1992.

[84] F. Pavese. Recalculation on ITS-90 of Accurate Vapour-Pressure Equations for e-H_2, Ne, N_2, O_2, Ar, CH_4, and CO_2. *J. Chem. Themodynamics*, 25:1351–1361, 1993.

[85] The document CCT_05_33 is available at http://www.bipm.org.

[86] E. S. Domalski. Selected Values of Heats of Combustion and Heats of Formation of Organic Compounds Containing the Elements C, H, N, O, P, and S. *J. Phys. Chem. Ref. Data*, 1:221–277, 1972.

[87] R. D. Freeman. Conversion of Standard (1 atm) Thermodynamic Data to the New Standard-State Pressure, 1 bar (10^5 Pa). *Bull. Chem. Thermodyn.*, 25:523–530, 1982.

[88] R. D. Freeman. Conversion of Standard (1 atm) Thermodynamic Data to the New Standard-State Pressure, 1 bar (10^5 Pa). *J. Chem. Eng. Data*, 29:105–111, 1984.

[89] R. D. Freeman. Conversion of Standard Thermodynamic Data to the New Standard-State Pressure. *J. Chem. Educ.*, 62:681–686, 1985.

[90] I. Tinoco Jr., K. Sauer, and J. C. Wang. *Physical Chemistry*, 4th edition. Prentice-Hall, New Jersey, 2001.

[91] R. A. Alberty. Recommendations for Nomenclature and Tables in Biochemical Thermodynamics. *Pure Appl. Chem.*, 66:1641–1666, 1994.

[92] D. D. Wagman, W. H. Evans, V. B. Parker, R. H. Schumm, I. Halow, S. M. Bailey, K. L. Churney, and R. L. Nuttall. The NBS Tables of Chemical Thermodynamic Properties. *J. Phys. Chem. Ref. Data, Vol. 11, Suppl. 2*, 1982.

[93] M. W. Chase Jr., C. A. Davies, J. R. Downey Jr., D. J. Frurip, R. A. McDonald, and A. N. Syverud. JANAF Thermochemical Tables, 3rd edition. *J. Phys. Chem. Ref. Data, Vol. 14, Suppl. 1*, 1985.

[94] V. P. Glushko, editor. *Termodinamicheskie Svoistva Individualnykh Veshchestv, Vols. 1-4.* Nauka, Moscow, 1978-85.

[95] I. Barin, editor. *Thermochemical Data of Pure Substances,* 3rd edition. VCH, Weinheim, 1995.

[96] M. W. Chase Jr. NIST JANAF Thermochemical Tables, 4th edition. *J. Phys. Chem. Ref. Data, Monograph 9, Suppl.,* 1998.

[97] J. P. Cali and K. N. Marsh. An Annotated Bibliography on Accuracy in Measurement. *Pure Appl. Chem.,* 55:907–930, 1983.

[98] G. Olofsson. Assignment and Presentation of Uncertainties of the Numerical Results of Thermodynamic Measurements. *Pure Appl. Chem.,* 53:1805–1825, 1981.

[99] CODATA Task Group on Data for Chemical Kinetics. The Presentation of Chemical Kinetics Data in the Primary Literature. *CODATA Bull.,* 13:1–7, 1974.

[100] K. J. Laidler. A Glossary of Terms Used in Chemical Kinetics, Including Reaction Dynamics. *Pure Appl. Chem.,* 68:149–192, 1996.

[101] M. L. Goldberger and K. M. Watson. *Collision Theory.* Krieger, Huntington (NY), 1975.

[102] M. Quack and J. Troe. Statistical Adiabatic Channel Model. In P. von Ragué Schleyer, N. Allinger, T. Clark, J. Gasteiger, P. A. Kollman, H. F. Schaefer III, and P. R. Schreiner, editors, *Encyclopedia of Computational Chemistry, Vol. 4,* pages 2708–2726, Chichester, UK, 1998. John Wiley and Sons.

[103] P. van Rysselberghe. Internationales Komitee für elektrochemische Thermodynamik und Kinetik CITCE. Bericht der Kommission für elektrochemische Nomenklatur und Definitionen. *Z. Elektrochem.,* 58:530–535, 1954.

[104] R. Parsons. Electrode Reaction Orders, Transfer Coefficients and Rate Constants. Amplification of Definitions and Recommendations for Publication of Parameters. *Pure Appl. Chem.,* 52:233–240, 1980.

[105] N. Ibl. Nomenclature for Transport Phenomena in Electrolytic Systems. *Pure Appl. Chem.,* 53:1827–1840, 1981.

[106] S. Trasatti. The Absolute Electrode Potential: An Explanatory Note. *Pure Appl. Chem.,* 58:955–966, 1986.

[107] A. J. Bard, R. Parsons, and J. Jordan, editors. *Standard Potentials in Aqueous Solutions.* Marcel Dekker Inc., New York, 1985.

[108] G. Gritzner and G. Kreysa. Nomenclature, Symbols, and Definitions in Electrochemical Engineering. *Pure Appl. Chem.,* 65:1009–1020, 1993.

[109] A. J. Bard, R. Memming, and B. Miller. Terminology in Semiconductor Electrochemistry and Photoelectrochemical Energy Conversion. *Pure Appl. Chem.,* 63:569–596, 1991.

[110] K. E. Heusler, D. Landolt, and S. Trasatti. Electrochemical Corrosion Nomenclature. *Pure Appl. Chem.,* 61:19–22, 1989.

[111] M. Sluyters-Rehbach. Impedances of Electrochemical Systems: Terminology, Nomenclature and Representation. Part 1: Cells with Metal Electrodes and Liquid Solutions. *Pure Appl. Chem.,* 66:1831–1891, 1994.

[112] A. V. Delgado, F. González-Caballero, R. J. Hunter, L. K. Koopal, and J. Lyklema. Measurement and Interpretation of Electrokinetic Phenomena. *Pure Appl. Chem.*, 77:1753–1805, 2005.

[113] D. R. Thévenot, K. Toth, R. A. Durst, and G. S. Wilson. Electrochemical Biosensors: Recommended Definitions and Classification. *Pure Appl. Chem.*, 71:2333–2348, 1999.

[114] R. A. Durst, A. J. Bäumner, R. W. Murray, R. P. Buck, and C. P. Andrieux. Chemically Modified Electrodes: Recommended Terminology and Definitions. *Pure Appl. Chem.*, 69:1317–1323, 1997.

[115] T. Cvitaš and I. M. Mills. Replacing Gram-Equivalents and Normalities. *Chem. Int.*, 16:123–124, 1994.

[116] R. P. Buck, S. Rondinini, A. K. Covington, F. G. K. Baucke, C. M. A. Brett, M. F. Camões, M. J. T. Milton, T. Mussini, R. Naumann, K. W. Pratt, P. Spitzer, and G. S. Wilson. Measurement of pH. Definition, Standards, and Procedures. *Pure Appl. Chem.*, 74:2169–2200, 2002.

[117] R. G. Bates and E. A. Guggenheim. Report on the Standardization of pH and Related Terminology. *Pure Appl. Chem.*, 1:163–168, 1960.

[118] R. G. Bates, editor. *Determination of pH – Theory and Practice*. John Wiley, New York, 1973.

[119] P. R. Mussini, T. Mussini, and S. Rondinini. Reference Value Standards and Primary Standards for pH Measurements in D_2O and Aqueous-Organic Solvent Mixtures: New Accessions and Assessments. *Pure Appl. Chem.*, 69:1007–1014, 1997.

[120] K. S. W. Sing, D. H. Everett, R. A. W. Haul, L. Moscou, R. A. Pierotti, J. Rouquérol, and T. Siemieniewska. Reporting Physisorption Data for Gas/Solid Systems with Special Reference to the Determination of Surface Area and Porosity. *Pure Appl. Chem.*, 57:603–619, 1985.

[121] L. Ter-Minassian-Saraga. Reporting Experimental Pressure-Area Data with Film Balances. *Pure Appl. Chem.*, 57:621–632, 1985.

[122] D. H. Everett. Reporting Data on Adsorption from Solution at the Solid/Solution Interface. *Pure Appl. Chem.*, 58:967–984, 1986.

[123] L. Ter-Minassian-Saraga. Thin Films Including Layers: Terminology in Relation to their Preparation and Characterization. *Pure Appl. Chem.*, 66:1667–1738, 1994.

[124] J. Haber. Manual on Catalyst Characterization. *Pure Appl. Chem.*, 63:1227–1246, 1991.

[125] W. V. Metanomski. *Compendium of Macromolecular Nomenclature*. Blackwell, Oxford, 1991.

[126] M. A. Van Hove, W. H. Weinberg, and C.-M. Chan. *Low-Energy Electron Diffraction: Experiment, Theory, and Surface Structure Determination*. Springer, Berlin, 1986.

[127] D. Kind and T. J. Quinn. Metrology: Quo Vadis? *Physics Today*, 52:13–15, 1999.

[128] I. M. Mills, P. J. Mohr, T. J. Quinn, B. N. Taylor, and E. R. Williams. Redefinition of the Kilogram: A Decision whose Time has come. *Metrologia*, 42:71–80, 2005.

[129] I. M. Mills. On Defining Base Units in Terms of Fundamental Constants. *Mol. Phys.*, 103:2989–2999, 2005.

[130] R. Dybkaer. Unit "katal" for Catalytic Activity. *Pure Appl. Chem.*, 73:927–931, 2001.

[131] I. Mills and C. Morfey. On Logarithmic Ratio Quantities and their Units. *Metrologia*, 42:246–252, 2005.

[132] H. G. Jerrard and D. B. McNeill. *A Dictionary of Scientific Units.* Chapman Hall, London, 3rd edition, 1980.

[133] W.-M. Yao, C. Amsler, D. Asner, R. M. Barnett, J. Beringer, P. R. Burchat, C. D. Carone, C. Caso, O. Dahl, G. D'Ambrosio, A. DeGouvea, M. Doser, S. Eidelman, J. L. Feng, T. Gherghetta, M. Goodman, C. Grab, D. E. Groom, A. Gurtu, K. Hagiwara, K. G. Hayes, J. J. Hernández-Rey, K. Hikasa, H. Jawahery, C. Kolda, Kwon Y., M. L. Mangano, A. V. Manohar, A. Masoni, R. Miquel, K. Mönig, H. Murayama, K. Nakamura, S. Navas, K. A. Olive, L. Pape, C. Patrignani, A. Piepke, G. Punzi, G. Raffelt, J. G. Smith, M. Tanabashi, J. Terning, N. A. Törnqvist, T. G. Trippe, P. Vogel, T. Watari, C. G. Wohl, R. L. Workman, P. A. Zyla, B. Armstrong, G. Harper, V. S. Lugovsky, P. Schaffner, M. Artuso, K. S. Babu, H. R. Band, E. Barberio, M. Battaglia, H. Bichsel, O. Biebel, P. Bloch, E. Blucher, R. N. Cahn, D. Casper, A. Cattai, A. Ceccucci, D. Chakraborty, R. S. Chivukula, G. Cowan, T. Damour, T. DeGrand, K. Desler, M. A. Dobbs, M. Drees, A. Edwards, D. A. Edwards, V. D. Elvira, J. Erler, V. V. Ezhela, W. Fetscher, B. D. Fields, B. Foster, D. Froidevaux, T. K. Gaisser, L. Garren, H.-J. Gerber, G. Gerbier, L. Gibbons, F. J. Gilman, G. F. Giudice, A. V. Gritsan, M. Grünewald, H. E. Haber, C. Hagmann, I. Hinchliffe, A. Höcker, P. Igo-Kemenes, J. D. Jackson, K. F. Johnson, D. Karlen, B. Kayser, D. Kirkby, S. R. Klein, K. Kleinknecht, I. G. Knowles, R. V. Kowalewski, P. Kreitz, B. Krusche, Yu. V. Kuyanov, O. Lahav, P. Langacker, A. Liddle, Z. Ligeti, T. M. Liss, L. Littenberg, L. Liu, K. S. Lugovsky, S. B. Lugovsky, T. Mannel, D. M. Manley, W. J. Marciano, A. D. Martin, D. Milstead, M. Narain, P. Nason, Y. Nir, J. A. Peacock, S. A. Prell, A. Quadt, S. Raby, B. N. Ratcliff, E. A. Razuvaev, B. Renk, P. Richardson, S. Roesler, G. Rolandi, M. T. Ronan, L. J. Rosenberg, C. T. Sachrajda, S. Sarkar, M. Schmitt, O. Schneider, D. Scott, T. Sjöstrand, G. F. Smoot, P. Sokolsky, S. Spanier, H. Spieler, A. Stahl, T. Stanev, R. E. Streitmatter, T. Sumiyoshi, N. P. Tkachenko, G. H. Trilling, G. Valencia, K. van Bibber, M. G. Vincter, D. R. Ward, B. R. Webber, J. D. Wells, M. Whalley, L. Wolfenstein, J. Womersley, C. L. Woody, A. Yamamoto, O. V. Zenin, J. Zhang, and R.-Y. Zhu. Review of Particle Physics, Particle Data Group. *J. Phys.*, G33:1–1232, 2006. The Particle Data Group website is http://pdg.lbl.gov.

[134] P. Seyfried and P. Becker. The Role of N_A in the SI: An Atomic Path to the Kilogram. *Metrologia*, 31:167–172, 1994.

[135] N. E. Holden. Atomic Weights of the Elements 1979. *Pure Appl. Chem.*, 52:2349–2384, 1980.

[136] P. De Bièvre and H. S. Peiser. 'Atomic Weight' – The Name, its History, Definition, and Units. *Pure Appl. Chem.*, 64:1535–1543, 1992.

[137] H. S. Peiser, N. E. Holden, P. De Bièvre, I. L. Barnes, R. Hagemann, J. R. de Laeter, T. J. Murphy, E. Roth, M. Shima, and H. G. Thode. Element by Element Review of their Atomic Weights. *Pure Appl. Chem.*, 56:695–768, 1984.

[138] K. J. R. Rosman and P. D. P. Taylor. Inorganic Chemistry Division, Commission on Atomic Weights and Isotopic Abundances, Subcommittee for Isotopic Abundance Measurements. Isotopic Compositions of the Elements 1997. *Pure Appl. Chem.*, 70:217–235, 1998.

[139] M. E. Wieser. Inorganic Chemistry Division, Commission on Isotopic Abundances and Atomic Weights. Atomic Weights of the Elements 2005. *Pure Appl. Chem.*, 78:2051–2066, 2006.

[140] W. C. Martin and W. L. Wiese. Atomic Spectroscopy. In G. W. F. Drake, editor, *Atomic, Molecular, and Optical Physics Reference Book*. American Institute of Physics, 1995.

[141] J. Corish and G. M. Rosenblatt. Name and Symbol of the Element with Atomic Number 110. *Pure Appl. Chem.*, 75:1613–1615, 2003.

[142] J. Corish and G. M. Rosenblatt. Name and Symbol of the Element with Atomic Number 111. *Pure Appl. Chem.*, 76:2101–2103, 2004.

[143] T. B. Coplen and H. S. Peiser. History of the Recommended Atomic-Weight Values from 1882 to 1997: A Comparison of Differences from Current Values to the Estimated Uncertainties of Earlier Values. *Pure Appl. Chem.*, 70:237–257, 1998.

[144] G. Audi, O. Bersillon, J. Blachot, and A. H. Wapstra. The NUBASE Evaluation of Nuclear and Decay Properties. *Nucl. Phys. A*, 729:3–128, 2003.

[145] A. H. Wapstra, G. Audi, and C. Thibault. The AME2003 Atomic Mass Evaluation. (I). Evaluation of Input Data, Adjustment Procedures. *Nucl. Phys. A*, 729:129–336, 2003.

[146] G. Audi, A. H. Wapstra, and C. Thibault. The AME2003 Atomic Mass Evaluation. (II). Tables, Graphs and References. *Nucl. Phys. A*, 729:337–676, 2003.

[147] J. K. Böhlke, J. R. de Laeter, P. de Bièvre, H. Hidaka, H. S. Peiser, K. J. R. Rosman, and P. D. P. Taylor. Inorganic Chemistry Division, Commission on Atomic Weights and Isotopic Abundances. Isotopic Compositions of the Elements, 2001. *J. Phys. Chem. Ref. Data*, 34:57–67, 2005.

[148] G. Audi and A. H. Wapstra. The 1993 Atomic Mass Evaluation. I. Atomic Mass Table. *Nucl. Phys. A*, 565:1–65, 1993.

[149] J. R. de Laeter, J. K. Böhlke, P. de Bièvre, H. Hidaka, H. S. Peiser, K. J. R. Rosman, and P. D. P. Taylor. Inorganic Chemistry Division, Commission on Atomic Weights and Isotopic Abundances. Atomic Weights of the Elements: Review 2000. *Pure Appl. Chem.*, 75:683–800, 2003.

[150] T. B. Coplen, J. K. Böhlke, P. De Bièvre, T. Ding, N. E. Holden, J. A. Hopple, H. R. Krouse, A. Lamberty, H. S. Peiser, K. Révész, S. E. Rieder, K. J. R. Rosman, E. Roth, P. D. P. Taylor, R. D. Vocke Jr., and Y. K. Xiao. Inorganic Chemistry Division, Commission on Atomic Weights and Isotopic Abundances, Subcommittee on Natural Isotopic Fractionation. Isotope-Abundance Variations of Selected Elements. *Pure Appl. Chem.*, 74:1987–2017, 2002.

[151] N. J. Stone. Table of Nuclear Magnetic Dipole and Electric Quadrupole Moments. *Atomic Data and Nuclear Data Tables*, 90:75–176, 2005.

[152] B. J. Jaramillo and R. Holdaway. *The Astronomical Almanac for the Year 2003*. U.S. Government Printing Office, Washington, 2003.

[153] R. T. Birge. On Electric and Magnetic Units and Dimensions. *The American Physics Teacher*, 2:41–48, 1934.

[154] J. D. Jackson. *Classical Electrodynamics*, 2nd edition. John Wiley, New York, 1975.

[155] E. A. Guggenheim. Names of Electrical Units. *Nature*, 148:751, 1941.

[156] D. R. Lide Jr. Use of Abbreviations in the Chemical Literature. *Pure Appl. Chem.*, 52:2229–2232, 1980.

[157] H. Q. Porter and D. W. Turner. A Descriptive Classification of the Electron Spectroscopies. *Pure Appl. Chem.*, 59:1343–1406, 1987.

[158] N. Sheppard. English-Derived Abbreviations for Experimental Techniques in Surface Science and Chemical Spectroscopy. *Pure Appl. Chem.*, 63:887–893, 1991.

[159] D. A. W. Wendisch. *Acronyms and Abbreviations in Molecular Spectroscopy*. Springer, Heidelberg, 1990.

11 GREEK ALPHABET

Roman	Italics	*Name*	*Pronounciation and Latin Equivalent*	*Notes*
A, α	*A, α*	alpha	A	
B, β	*B, β*	beta	B	
Γ, γ	*Γ, γ*	gamma	G	
Δ, δ	*Δ, δ*	delta	D	
E, ε	*E, ε, ε*	epsilon	E	
Z, ζ	*Z, ζ*	zeta	Z	
H, η	*H, η*	eta	Ae, Ä, Ee	1
Θ, ϑ, θ	*Θ, ϑ, θ*	theta	Th	2
I, ι	*I, ι*	iota	I	
K, ϰ, κ	*K, ϰ, κ*	kappa	K	2
Λ, λ	*Λ, λ*	lambda	L	
M, μ	*M, μ*	mu, (my)	M	
N, ν	*N, ν*	nu, (ny)	N	
Ξ, ξ	*Ξ, ξ*	xi	X	
O, o	*O, o*	omikron	O	
Π, π	*Π, π*	pi	P	
P, ρ	*P, ρ*	rho	R	
Σ, σ, ς	*Σ, σ, ς*	sigma	S	2, 3
T, τ	*T, τ*	tau	T	
Y, υ	*Y, υ*	upsilon, ypsilon	U, Y	
Φ, φ, ϕ	*Φ, φ, ϕ*	phi	Ph	2
X, χ	*X, χ*	chi	Ch	
Ψ, ψ	*Ψ, ψ*	psi	Ps	
Ω, ω	*Ω, ω*	omega	Oo	4

(1) For the Latin equivalent Ae is to be pronounced as the German Ä. The modern Greek pronounciation of the letter η is like E, long ee as in cheese, or short i as in lips. The Latin equivalent is also often called "long E".

(2) For the lower case letters theta, kappa, sigma and phi there are two variants in each case. For instance, the second variant of the lower case theta is sometimes called "vartheta" in printing.

(3) The second variant for lower case sigma is used in Greek only at the end of the word.

(4) In contrast to omikron (short o) the letter omega is pronounced like a long o.

12 INDEX OF SYMBOLS

This index lists symbols for physical quantities, units, some selected mathematical operators, states of aggregation, processes, and particles. Symbols for chemical elements are given in Section 6.2, p. 117. Qualifying subscripts and superscripts, etc., are generally omitted from this index, so that for example E_p for potential energy, and E_{ea} for electron affinity are both indexed under E for energy. The Latin alphabet is indexed ahead of the Greek alphabet, lower case letters ahead of upper case, bold ahead of italic, ahead of Roman, and single letter symbols ahead of multiletter ones. When more than one page reference is given, bold print is used to indicate the general (defining) reference. Numerical entries for the corresponding quantities are underlined.

183

184

	<u>111</u>, 134, 143–146
ε	Levi-Civita symbol, 106
ε	unit step function, Heaviside function, 45, 67, **107**
ζ	Coriolis ζ-constant, **26**, 30
ζ	electrokinetic potential, 73
ζ	magnetizability, 23
ζ	shielding parameter, 21
η	nuclear Overhauser enhancement, 29
η	overpotential, 72
η	viscosity, **15**, 81, 82, 138
ϑ	cylindrical coordinate, 13
ϑ	plane angle, **13**, 38, 89, 92, 136
ϑ	volume strain, bulk strain, 15
θ	Bragg angle, 42
θ	Celsius temperature, 38, 39, 41, 46, **56**, 61–64, 87, 129, 133
θ	characteristic temperature, 46
θ	characteristic (Weiss) temperature, 43
θ	contact angle, 77
θ	scattering angle, 65
θ	spherical polar coordinate, **13**, 18, 21, 96
θ	surface coverage, 77, 80
θ	internal vibrational coordinate, 27
θ_{W}	characteristic (Weiss) temperature, 43
θ_{W}	weak mixing angle, Weinberg angle, **24**, <u>111</u>, 112
Θ	quadrupole moment, 23, **24**, 141
Θ	characteristic temperature, 46
Θ_{D}	Debye temperature, **43**, 80
Θ	plane angle, 35
Θ	temperature, 56
κ	asymmetry parameter, 25
κ	bulk viscosity, 81
κ	compressibility, 43, **56**
κ	conductivity, 6, **17**, 73, 81, 82, 90, 132, 140, 147
κ	magnetic susceptibility, **17**, 133, 142, 145, 147
κ	molar napierian absorption coefficient, **36**, 37
κ	ratio of heat capacities, 57
κ	transmission coefficient, 65, **66**
λ	thermal conductivity tensor, 43
λ	absolute activity, 46, **57**, 58
λ	angular momentum quantum number (component), 30
λ	decay constant, 24
λ	mean free path, 44, **65**, 80, 81

λ	molar conductivity of an ion, 54, **73**
λ	thermal conductivity, 43, 44, **81**, 90
λ	van der Waals constant, 77
λ	wavelength, 3, 34–38, 42–44, 131
λ	lambda (unit of volume), <u>136</u>
Λ	angular momentum quantum number (component), 30–32
Λ	molar conductivity, 6, 54, **73**, 90, 132
$\boldsymbol{\mu}$	electric dipole moment, 17, 23, **26**, 35, 39, 41, 95, 141, 146
μ	chemical potential, 5, 6, 8, 43, 44, 46, **57**, 58, 60–62
μ	electric mobility, 73
$\tilde{\mu}$	electrochemical potential, 71
μ	dynamic friction factor, 15
μ	Joule-Thomson coefficient, 57
μ	magnetic dipole moment, 17, **23**, 95, 115, 121, 133, 142
μ	mean, 151
μ	mobility, 43
μ	permeability, 5, **17**, 81, 90
μ	reduced mass, **14**, 64–66, 95
μ	Thomson coefficient, 43
μ	viscosity, **15**, 81
$\tilde{\mu}$	electrochemical potential, 71
μ_0	permeability of vacuum, magnetic constant, 16, **17**, 28, <u>111</u>, 133, 143, 144, 147
μ_{B}	Bohr magneton, **23**, 27, 29, 95, <u>111</u>, 112, 115, 133, 142
μ_{e}	electron magnetic moment, **23**, <u>111</u>
μ_{N}	nuclear magneton, **23**, 29, <u>111</u>, 115, 116, 121, 142
μ_{p}	proton magnetic moment, <u>112</u>, 116
μ	micro (SI prefix), 91
μ	micron (unit of length), <u>135</u>
μ	muon, 8, **50**, <u>115</u>, 116
$\boldsymbol{\nu}$	stoichiometric number matrix, 53
ν	charge number of cell reaction, 71
ν_{D}	Debye frequency, 43
ν	frequency, **13**, 23, 25, 28, 29, 33–36, 41, 68, 81, 88, 89, 129
ν	kinematic viscosity, **15**, 90, 138
ν	stoichiometric number, **53**, 58, 61, 63, 71–74
$\tilde{\nu}_{\mathrm{D}}$	Debye wavenumber, 43
$\tilde{\nu}$	wavenumber (in vacuum), **25**, 26, 34–39, 131
ν	vibrational state symbol, 33
ν_{e}	electron neutrino, **50**, <u>115</u>
ξ	Coriolis coupling constant, 26
ξ	extent of reaction, advancement, **48**, 49, <u>53</u>, 58, 60, 63, 67

192

ω	angular frequency, angular velocity, 8, **13**, 23, 34, 89, 98
ω_D	Debye angular frequency, 43
ω	degeneracy, statistical weight, 26, **45**
ω	harmonic (vibrational) wavenumber, 25, 26
ω	solid angle, **13**, 34, 36, 65, 89
Ω	angular momentum quantum number (component), 30, 32
Ω	nutation angular frequency, 28
Ω	partition function, 46
Ω	solid angle, **13**, 34–36, 65, 89
Ω	volume in phase space, 45
Ω	ohm, **89**, <u>140</u>

Special symbols

%	percent, 97, 98
‰	permille, 97, 98
°	degree (unit of arc), 38, **92**, <u>136</u>
⊖, °	standard (superscript), **60**, 71
′	minute (unit of arc), **92**, <u>136</u>
″	second (unit of arc), **92**, <u>136</u>
*	complex conjugate, 107
*	excitation, 50
*	pure substance (superscript), 60
‡, ≠	activated complex (superscript), transition state (superscript), 60
∞	infinite dilution (superscript), 60
e	even parity symmetry label, 32
o	odd parity symmetry label, 32
[B]	amount concentration, concentration of B, **48**, 54, 69
Δ_r	derivative with respect to extent of reaction, **58**, 60, 74
$\dim(Q)$	dimension of quantity Q, 4
∇	nabla, 16–18, 20, 43, 81, 96, **107**, 146, 147
$[\alpha]_\lambda^\theta$	specific optical rotatory power, **37**, 38
$[Q]$	unit of quantity Q, 4

13 SUBJECT INDEX

When more than one page reference is given, bold print is used to indicate the general (defining) reference. Underlining is used to indicate a numerical entry of the corresponding physical quantity. Greek letters are ordered according to their spelling and accents are ignored in alphabetical ordering. Plural form is listed as singular form, where applicable.

197

astronomical day, 137
astronomical unit, 9, **92**, <u>94</u>, <u>136</u>
astronomy, 94
asymmetric reduction, 25
asymmetric top, **30**, 31, 33
asymmetry-doubling, 31
asymmetry parameter, 25
    Wang, 25
atmosphere, 131
    litre, <u>137</u>
    standard, **62**, <u>111</u>, <u>138</u>
atmospheric pressure, 58
atomic absorption spectroscopy, 97
atomic mass, **22**, 47, 121
    average relative, 120
    relative, **47**, 117–120
atomic mass constant, **22**, 47, <u>111</u>, 117
atomic mass evaluation, 121
atomic mass unit
    unified, 9, 22, 47, **92**, 94, <u>111</u>, 116, 117, 121, 136
atomic number, **22**, 23, 49, 50, 117, 121
atomic orbital, 19
atomic-orbital basis function, **19**, 21
atomic physics, 22
atomic scattering factor, 42
atomic spectroscopy, 14
atomic standard, 137
atomic state, 32
atomic unit, 4, 18, 20, 22, 24, 26, **94**, <u>95</u>, 96, 129, 132, 135, 143–145
    symbol for, 94–96
atomic unit of magnetic dipole moment, 95
atomic units
    system of, 145
atomic weight, **47**, 117–120
atomization, 61
    energy of, 61
atomization (subscript), 59
atto, 91
au, 139
au of action, <u>138</u>
au of force, <u>137</u>
au of time, <u>137</u>
average, 41, **49**
average collision cross section, 64
averaged collision cross section, 64, **65**
average distance
    zero-point, 27
average lifetime, 64
average mass, 117
average molar mass, 77
average molar mass (z-average), 77
average molar mass (mass-average), 77
average molar mass (number-average), 77
average population, 46
average probability, 46

average relative atomic mass, 120
average speed, **45**, 81
Avogadro constant, 4, 9, **45**, 47, 53, <u>111</u>, 112
avoirdupois, <u>136</u>
axis of quantization, 30
azimuthal quantum number, 30

band gap energy, 43
band intensity, 26, 38–40
bar, 48, **131**, <u>138</u>, 233
barn, **121**, <u>136</u>
barrel (US), <u>136</u>
base-centred lattice, 44
base hydrolysis, 59
base of natural logarithm, 8
base quantity, **4**, 16, 85, 86
    symbol of, 4
base unit, 34, 85–89, 93, 135, 143, 144
    SI, 4, 85–88, 90, 91
basis function
    atomic-orbital, **19**, 21
Bates-Guggenheim convention, 75
becquerel, **89**, <u>139</u>
Beer-Lambert law, **35**, 36
bel, **92**, 98, 99
bending coordinate, 27
best estimate, 149
β-particle, 50, <u>115</u>
billion, 98
bimolecular reaction, **68**, 69
binary multiple, 91
binominal coefficient, 105
biochemical standard state, 62
biot, **135**, <u>139</u>, 144
Biot-Savart law, 146
BIPM, **5**, 9
black-body radiation, 87
black body, 35
Bloch function, 43
body-centred lattice, 44
bohr, <u>135</u>, 145
Bohr magneton, **23**, 29, 95, <u>111</u>, 115, 133, 142
Bohr radius, 9, **22**, 95, <u>111</u>
Boltzmann constant, 36, **45**, 46, 64, 81, <u>111</u>
bond dissociation energy, 22
bonding interaction, 19
bond length
    equilibrium, 96
bond order, **19**, 21
boson
    W-, 111, <u>115</u>
    Z-, 111, <u>115</u>
Bragg angle, 42
Bragg reflection, 44
branches (rotational band), 33
Bravais lattice vector, 42
breadth, 13

British thermal unit, <u>137</u>
bulk lattice, 79
bulk modulus, 15
bulk phase, **40**, 71, 77
bulk plane, 79
bulk strain, 15
bulk viscosity, 81
Burgers vector, 42

calorie
    15 °C, <u>137</u>
    international, <u>137</u>
    thermochemical, **58**, <u>137</u>
candela, 34, 85, 86, **88**
canonical ensemble, **45**, 46
capacitance, **16**, <u>140</u>, 146
    electric, 89
cartesian axis
    space-fixed, **35**, 39
cartesian component, 107
cartesian coordinate system, 107
cartesian (space) coordinate, **13**, 21
catalytic activity, 89
catalytic reaction mechanism, 68
cathode, 75
cathodic current, 70
cathodic partial current, 72
cathodic transfer coefficient, 72
cavity, 36
    Fabry-Perot, 36
    laser, 36
CAWIA, 117, 121
cell
    electrochemical, 52, 71, 73–75
    electrolytic, **73**, 75
cell diagram, 71, 72, **73**, 74
cell potential, 16, **71**, 89
cell reaction
    electrochemical, 52, **74**
cell without transference, 75
Celsius scale
    zero of, <u>111</u>
Celsius temperature, 38, 46, **56**, 89, 111, <u>138</u>
centi, 91
centigrade temperature, 56
centimetre, 134, <u>135</u>, 144
centipoise, <u>138</u>
central coulomb field, 23
centre of mass, 31
centrifugal distortion constant, 25
centrifugation, 78
CGPM, xi, **85**, 87, 88, 92
CGS unit, 96, 135–138
chain-reaction mechanism, 68
characteristic Debye length, 78
characteristic impedance, <u>111</u>
characteristic number, 5

transport, 82
characteristic temperature, 46
characteristic terrestrial isotopic composition, 117, 120
characteristic time interval, 13
characteristic (Weiss) temperature, 43
charge, **16**, 18, 20, 43, 71, 72, 95, 115, 116
    areic, 6
    effective, 21
    electric, 4, **16**, 50, 70, 89, 134, 135, 139, 144
    electron, 18
    electronic, 19
    electrostatic unit of, 135
    electroweak, 24
    elementary, 8, 9, **22**, 24, 26, 70, 94–96, <u>111</u>, 145
    ion, 73
    point of zero, 78
    proton, **22**, 70
    reduced, 95
charge (au)
    proton, 139
charge density, **16**, 18, 23, 43, 44, 140
    electric, 90
    surface, 6, **71**
charge number, 20, 49–51, 70, 71, 73
    ionic, 49
charge number of cell reaction, 71
charge order, 19
charge partition, 74
charge transfer, **71**, 72, 74
chemical amount, 4, 45, **47**
chemical composition, 47, **51**, 71
chemical-compound name, 8
chemical element, 49
    symbol for, 8, 9, **49**, 50, 113, 117–120
chemical energy, 73
chemical equation, 49, **53**
    bimolecular elementary reaction, 68
    elementary reaction, **68**, 69
    general, 53
    stoichiometric, **52**, **60**, 63, 68, 69
    unimolecular elementary reaction, 68
chemical equilibrium, 71, **74**
    local, 72
chemical formula, 49–51
chemical kinetics, **63**, 68
chemical potential, 5, 6, 8, 44, 46, **57**, 62
    standard, **57**, 61, 62
chemical reaction, **53**, 60, 63, 73
    equation for, 52
    transition state theory, 65
chemical shift, 29
chemical substance, 51
chemical symbol for element, 50
chemical thermodynamics, 56–62
chemically modified electrode, 70

exchange current density, 72
exchange integral, 20
exchange operator, **20**, 21
excitation, 50
excitation energy, **131**, 132
  molar, 132
excitation symbol, 50
excited-state label, 33
excited electronic state, 8, **50**
excited state, 32
excited state concentration, 67
expanded uncertainty, 151–153
expansion coefficient
  cubic, 43, **56**, 81
  linear, 56
expansivity coefficient, 57
expectation value, **23**, 24
expectation value (core hamiltonian), 20
expectation value (operator), 18
expected best estimate, 149
expected dispersion, 149
expected value, 151
exponential decay, 67
exponential function, 106
exposure, 90
extensive, 6
extensive quantity, **6**, 57, 81
extensive quantity (divided by area), 81
extensive quantity (divided by volume), 81
extensive thermodynamic quantity, 60
extent of reaction, **48**, 53, 60
extinction, 37
extinction coefficient, 37

f-value, 37
Fabry-Perot cavity, 36
face-centred lattice, 44
factor
  absorption, 36
  acoustic, 15
  (acoustic) absorption, 15
  (acoustic) dissipation, 15
  (acoustic) reflection, 15
  (acoustic) transmission, 15
  atomic scattering, 42
  collision frequency, 65
  compressibility, 57
  compression, 57
  Debye-Waller, 42
  dynamic friction, 15
  frequency, 64
  normalization, **18**, 20, 21
  pre-exponential, 63, **64**
  quality, 36
  (radiation) absorption, 15
  reflection, 36
  structure, 42

symmetry, 72
transmission, 36
factorial, 105
Fahrenheit temperature, <u>138</u>
farad, 89
faradaic current, 72
Faraday constant, **70**, <u>111</u>
Faraday induction law, 147
femto, 91
femtometre, 121
fermi, <u>135</u>
Fermi coupling constant, <u>111</u>
Fermi energy, 43
Fermi sphere, 44
field, 98
  electric, **24**, 95, 104, 146
  magnetic, **17**, 29, 141
field gradient
  electric, **29**, 141
field gradient tensor, 29
field level, **98**, 99
field strength
  electric, 7, **16**, 24, 26, 73, 90, 134, 140, 145–147
  magnetic, **17**, 90, 142, 147
film, 78
film tension, 77
film thickness, 77
fine-structure constant, **22**, 95, <u>111</u>, 144, 145
finesse, 36
first-order kinetics
  generalized, 67
first-order rate constant, **66**, 67
first-order reaction, 72
Fischer projection, **51**, 52
flow, 81
flow rate, 81
fluence, 35
fluence rate, 35
fluidity, 15
fluid phase, 54
fluorescence, 67
fluorescence rate constant, **66**
flux, 72, 81
  electric, 16
  heat, 81
  luminous, 89
  magnetic, **17**, 89, 142
  probability, 18
  radiant, 89
  reflected, 15
  sound energy, 15
  transmitted, 15
flux density, 81
  effective magnetic, 29
  heat, **81**, 90
  magnetic, **17**, 28, 81, 89, 95, 141, 145–147

208

HMO, *see* Hückel (molecular orbital) theory
horse power
    imperial, 138
    metric, 138
hot-band transition, 41
hour, **92**, 137
Hückel (molecular orbital) theory, 19
Hückel secular determinant, 19
hydrogen-like atom, 23
hydrogen-like wavefunction, 18
hyper-polarizability, 17, **24**, 95
hyper-susceptibility, 16
hyperbolic function, 106
hyperfine coupling constant, 27
hyperfine coupling hamiltonian, 27
hyperfine level, 87

ideal (superscript), 60
ideal gas, 39, **61**, 66
    molar volume of, 111
identity operator, 31
IEC, 6, **91**, 92, 136
illuminance, 89
imaginary refractive index, **37**, 40
immersion (subscript), 60
impact parameter, 65
impedance, **17**, 70, 98
    characteristic, 111
    complex, 17
imperial horse power, 138
inch, 135, 138
incident intensity, 38
incoming particle, 50
indirect mechanism, 68
indirect spin coupling tensor, 28
individual collision, 65
induced emission, **35**, 37–39
induced radiative transition, 38
inductance, **17**, 89, 142
    mutual, 17
    self-, **17**, 142, 147
induction
    magnetic, **17**, 27
inertial defect, 25
inexact differential, 56
infinite dilution, **54**, 55, 61
infinite dilution (superscript), 60
infinitesimal electromagnetic force, 143
infinitesimal variation, 8
infrared spectra, 40
inner electric potential, 70
inner potential, 71
inner potential difference, 71
integral
    coulomb, **19**, 20
    exchange, 20
    inter-electron repulsion, 20

one-electron, **20**, 21
overlap, 19
resonance, 19
two-electron, 21
two-electron repulsion, 19, **20**
integrated absorption coefficient, **37**, 39–41
integrated absorption cross section, 38
integrated cross section, 39
integrated net absorption cross section, 37
integration element, 18
intensity, 34–36, 38
    absorption, 38–40
    band, 26, 38–40
    incident, 38
    line, 26, 38–40
    luminous, 4, **34**, 85, 86, 88
    photon, 34
    radiant, **34**, 35, 88
    signal, 29
    spectral, **35**, 36
    transmitted, 38
intensity of electromagnetic radiation, 35
intensive, **6**, 81
interaction energy
    electrostatic, 43
interaction tensor
    dipolar, 28
interatomic equilibrium distance, 27
interatomic (internuclear) distance, 27
inter-electron repulsion integral, 20
interface, 71
interfacial entropy per area, 78
interfacial layer, **77**, 78
interfacial tension, **77**, 78
internal absorptance, 36
internal coordinate, 27
internal energy, 55, **56**
internal energy of activation
    standard, 65
internal energy of atomization
    standard, 61
internal vibrational angular momentum, 30
internal vibrational coordinate, 27
international calorie, 137
international ohm
    mean, 140
    US, 140
international protype of the kilogram, 87
International System of Units (SI), 85
International temperature, 56
international volt
    mean, 140
    US, 140
intrinsic number density, 44
inverse of a square matrix, 107
inversion operation, **31**, 44

217

220

siemens, 89
sievert, **89**, <u>139</u>
signal amplitude level, 98
signal intensity, 29
signal power level, 98
signal transmission, 98
single-crystal surface, 79
SI prefix, 85, **91**, 92, 94
SI unit, 96
SI unit system, 143
Slater determinant, 20
Slater-type orbital, 21
solid, 27, 28, 38, 40, **54**, 79
    amorphous, **54**, 55
    crystalline, 55
solid absorbent, 77
solid angle, **13**, 34, 36, 65, 89
solid phase, 55, 59, **61**
solid state, 42–44, 61
solubility, **48**, 49
solubility product, 59
solute, 40, 48, 59, **61**, 62, 133
solution, 38, 40, 48, 52, 54, 57, **61**, 72, 74, 75, 132, 133
    aqueous, **54**, 55, 57, 74, 97
    dilute, **58**, 97
    molal, 48
solution (subscript), 60
solution at infinite dilution
    aqueous, **54**, 55
solvent, 29, 37, 48, 59, **61**, 133
    pure, 132
solvent mixture, 76
sound
    speed of, **13**, 81
sound energy flux, 15
sound power level, 99
sound pressure level, 99
space-fixed cartesian axis, **35**, 39
space-fixed component, **30**, 35
space-fixed coordinate, 31
space-fixed inversion operator, 31
space-fixed origin, 31
space and time, 13
space charge, 78
space coordinate
    cartesian, **13**, 21
specific, 6
specific conductance, 73
specific energy, 90
specific entropy, 90
specific heat capacity, 90
specific heat capacity (constant pressure), **6**, 81
specific number density, 44
specific optical rotatory power, **37**, 38
specific quantity, **6**, 56

specific rate constant, 66
specific rotation, 38
specific rotatory power, 38
specific surface area, 77
specific volume, **6**, 14, 90
spectral absorption, 38
spectral density of vibrational modes, 43
spectral intensity, **35**, 36
spectral irradiance, 35
spectral line, 37
spectral radiant energy density, **34**, 35
spectroscopic absorption intensity, 40
spectroscopic transition, 33
spectroscopy, 25–33, 36
    atomic, 14
    atomic absorption, 97
    molecular, 14
    optical, 32, 36
    spin-resonance, 23
spectrum
    absorption, 36
speed, **13**, 81, 90, 93, 95
    average, **45**, 81
    mean relative, 64
speed distribution function, 45
speed of light, 9, 13, 16, 25, **34**, 43, 95, <u>111</u>, 143–145
speed of sound, **13**, 81
spherical harmonic function, 18
spherical polar coordinate, **13**, 21
spherical symmetry, 145
spherical top, 30
spin
    nuclear, 29
spin angular momentum, **19**, 28, 116
spin-lattice relaxation time, 29
    longitudinal, 29
spin multiplicity, 32
    electron, 32
spin operator
    electron, 25, **27**
    nuclear, 28
spin-orbit coupling constant, 25
spin quantum number
    nuclear, 121
spin-resonance spectroscopy, 23
spin-rotation coupling constant, 28
spin-rotation interaction tensor, 28
spin-spin coupling constant
    nuclear, 28
    reduced nuclear, 28
spin-spin relaxation time, 29
spin statistical weight, 45
spin wavefunction, **19**, 29
    electron, 19
spontaneous emission, **35**, 67
square metre, <u>136</u>

224

NOTES

NOTES

PRESSURE CONVERSION FACTORS

	Pa	kPa	bar	atm	Torr	psi
1 Pa	$= 1$	$- 10^{-3}$	$- 10^{-5}$	$\approx 9.869\ 23\times10^{-6}$	$\approx 7.500\ 62\times10^{-3}$	$\approx 1.450\ 38\times10^{-4}$
1 kPa	$= 10^{3}$	$= 1$	$- 10^{-2}$	$\approx 9.869\ 23\times10^{-3}$	$\approx 7.500\ 62$	$\approx 0.145\ 038$
1 bar	$= 10^{5}$	$= 10^{2}$	$= 1$	$\approx 0.986\ 923$	≈ 750.062	≈ 14.5038
1 atm	$= 101\ 325$	$= 101.325$	$= 1.013\ 25$	$= 1$	$= 760$	≈ 14.6959
1 Torr	≈ 133.322	$\approx 0.133\ 322$	$\approx 1.333\ 22\times10^{-3}$	$\approx 1.315\ 79\times10^{-3}$	$= 1$	$\approx 1.933\ 68\times10^{-2}$
1 psi	≈ 6894.76	$\approx 6.894\ 76$	$\approx 6.894\ 76\times10^{-2}$	$\approx 6.804\ 60\times10^{-2}$	$\approx 51.714\ 94$	1

Examples of the use of this table: 1 bar $\approx 0.986\ 923$ atm

1 Torr ≈ 133.322 Pa

Note: 1 mmHg $= 1$ Torr, to better than 2×10^{-7} Torr, see Section 7.2, p. 138.

NUMERICAL ENERGY CONVERSION FACTORS

$$E = h\nu = hc\tilde{\nu} = kT; \quad E_m = N_A E$$

		wavenumber $\tilde{\nu}$ / cm⁻¹	frequency ν / MHz	energy E — aJ	eV	E_h	molar energy E_m — kJ mol⁻¹	kcal mol⁻¹	temperature T / K
$\tilde{\nu}$:	1 cm⁻¹ ≙	1	$2.997\,925\times10^{4}$	$1.986\,446\times10^{-5}$	$1.239\,842\times10^{-4}$	$4.556\,335\times10^{-6}$	$11.962\,66\times10^{-3}$	$2.859\,144\times10^{-3}$	$1.438\,775$
ν:	1 MHz ≙	$3.335\,641\times10^{-5}$	1	$6.626\,069\times10^{-10}$	$4.135\,667\times10^{-9}$	$1.519\,830\times10^{-10}$	$3.990\,313\times10^{-7}$	$9.537\,076\times10^{-8}$	$4.799\,237\times10^{-5}$
E:	1 aJ ≙	$50\,341.17$	$1.509\,190\times10^{9}$	1	$6.241\,510$	$0.229\,371\,3$	602.2142	143.9326	$7.242\,963\times10^{4}$
	1 eV ≙	8065.545	$2.417\,989\times10^{8}$	$0.160\,217\,6$	1	$3.674\,933\times10^{-2}$	$96.485\,34$	$23.060\,55$	$1.160\,451\times10^{4}$
	1 E_h ≙	$219\,474.6$	$6.579\,684\times10^{9}$	$4.359\,744$	$27.211\,38$	1	2625.500	627.5095	$3.157\,747\times10^{5}$
E_m:	1 kJ mol⁻¹ ≙	$83.593\,47$	$2.506\,069\times10^{6}$	$1.660\,539\times10^{-3}$	$1.036\,427\times10^{-2}$	$3.808\,799\times10^{-4}$	1	$0.239\,005\,7$	120.2722
	1 kcal mol⁻¹ ≙	349.7551	$1.048\,539\times10^{7}$	$6.947\,694\times10^{-3}$	$4.336\,410\times10^{-2}$	$1.593\,601\times10^{-3}$	4.184	1	503.2189
T:	1 K ≙	$0.695\,035\,6$	$2.083\,664\times10^{4}$	$1.380\,650\times10^{-5}$	$8.617\,343\times10^{-5}$	$3.166\,815\times10^{-6}$	$8.314\,472\times10^{-3}$	$1.987\,207\times10^{-3}$	1

The symbol ≙ should be read as meaning 'approximately corresponding to' or 'is approximately equivalent to'. The conversion from kJ to kcal is exact by definition of the thermochemical kcal (see Section 2.11, note 16, p. 58). The values in this table have been obtained from the constants in Chapter 5, p. 111. The last digit is given but may not be significant.

Examples of the use of this table:

1 aJ ≙ 50 341.17 cm⁻¹

1 eV ≙ 96.485 34 kJ mol⁻¹

Examples of the derivation of the conversion factors:

1 aJ to MHz

$$\frac{(1\,\text{aJ})}{h} \;\hat{=}\; \frac{10^{-18}\,\text{J}}{6.626\,068\,96 \times 10^{-34}\,\text{J s}} \;\hat{=}\; 1.509\,190 \times 10^{15}\,\text{s}^{-1} \;\hat{=}\; 1.509\,190 \times 10^{9}\,\text{MHz}$$

1 cm⁻¹ to eV

$$(1\,\text{cm}^{-1})\,hc\left(\frac{e}{e}\right) \;\hat{=}\; \frac{(1.986\,445\,501 \times 10^{-25}\,\text{J}) \times 10^{2}\,e}{1.602\,176\,487 \times 10^{-19}\,\text{C}} \;\hat{=}\; 1.239\,842 \times 10^{-4}\,\text{eV}$$

1 E_h to kJ mol⁻¹

$$(1\,E_h)\,N_A \;\hat{=}\; (4.359\,743\,94 \times 10^{-18}\,\text{J}) \times (6.022\,141\,79 \times 10^{23}\,\text{mol}^{-1}) \;\hat{=}\; 2625.500\,\text{kJ mol}^{-1}$$

1 kcal mol⁻¹ to cm⁻¹

$$\frac{(1\,\text{kcal mol}^{-1})}{hcN_A} \;\hat{=}\; \frac{4.184 \times (1\,\text{kJ mol}^{-1})}{hcN_A} \;\hat{=}\; \frac{4.184 \times (10^{3}\,\text{J mol}^{-1})}{(1.986\,445\,501 \times 10^{-25}\,\text{J}) \times 10^{2}\,\text{cm} \times (6.022\,141\,79 \times 10^{23}\,\text{mol}^{-1})} \;\hat{=}\; 349.7551\,\text{cm}^{-1}$$